Journal of Applied Logics - IfCoLog Journal of Logics and their Applications

Volume 12, Number 1

January 2025

Disclaimer

Statements of fact and opinion in the articles in Journal of Applied Logics - IfCoLog Journal of Logics and their Applications (JALs-FLAP) are those of the respective authors and contributors and not of the JALs-FLAP. Neither College Publications nor the JALs-FLAP make any representation, express or implied, in respect of the accuracy of the material in this journal and cannot accept any legal responsibility or liability for any errors or omissions that may be made. The reader should make his/her own evaluation as to the appropriateness or otherwise of any experimental technique described.

© Individual authors and College Publications 2025
All rights reserved.

ISBN 978-1-84890-478-1
ISSN (E) 2631-9829
ISSN (P) 2631-9810

College Publications
Scientific Director: Dov Gabbay
Managing Director: Jane Spurr

http://www.collegepublications.co.uk

All rights reserved. No part of this publication may be used for commercial purposes or transmitted in modified form by any means, electronic, mechanical, photocopying, recording or otherwise without prior permission, in writing, from the publisher.

Editorial Board

Editors-in-Chief
Dov M. Gabbay and Jörg Siekmann

Marcello D'Agostino
Natasha Alechina
Sandra Alves
Jan Broersen
Martin Caminada
Balder ten Cate
Agata Ciabattoni
Robin Cooper
Luis Farinas del Cerro
Esther David
Didier Dubois
PM Dung
David Fernandez Duque
Jan van Eijck
Marcelo Falappa
Amy Felty
Eduaro Fermé
Melvin Fitting

Michael Gabbay
Murdoch Gabbay
Thomas F. Gordon
Wesley H. Holliday
Sara Kalvala
Shalom Lappin
Beishui Liao
David Makinson
Réka Markovich
George Metcalfe
Claudia Nalon
Valeria de Paiva
Jeff Paris
David Pearce
Pavlos Peppas
Brigitte Pientka
Elaine Pimentel
Henri Prade

David Pym
Ruy de Queiroz
Ram Ramanujam
Christian Retoré
Ulrike Sattler
Jörg Siekmann
Marija Slavkovik
Jane Spurr
Kaile Su
Leon van der Torre
Yde Venema
Rineke Verbrugge
Jun Tao Wang
Heinrich Wansing
Jef Wijsen
Michael Wooldridge
Anna Zamansky

Scope and Submissions

This journal considers submission in all areas of pure and applied logic, including:

- pure logical systems
- proof theory
- constructive logic
- categorical logic
- modal and temporal logic
- model theory
- recursion theory
- type theory
- nominal theory
- nonclassical logics
- nonmonotonic logic
- numerical and uncertainty reasoning
- logic and AI
- foundations of logic programming
- belief change/revision
- systems of knowledge and belief
- logics and semantics of programming
- specification and verification
- agent theory
- databases
- dynamic logic
- quantum logic
- algebraic logic
- logic and cognition
- probabilistic logic
- logic and networks
- neuro-logical systems
- complexity
- argumentation theory
- logic and computation
- logic and language
- logic engineering
- knowledge-based systems
- automated reasoning
- knowledge representation
- logic in hardware and VLSI
- natural language
- concurrent computation
- planning

This journal will also consider papers on the application of logic in other subject areas: philosophy, cognitive science, physics etc. provided they have some formal content.

Submissions should be sent to Jane Spurr (jane@janespurr.net) as a pdf file, preferably compiled in LaTeX using the IFCoLog class file.

CONTENTS

ARTICLES

Editorial 1
   Martin Lukac and Mike Behrisch

Multiple-Valued Reversible Fredkin Gates and Ensembles 3
   Claudio Moraga

Alternative Cut-free Sequent Calculi for S4 that are Compatible with
   Paradefinite Four-valued Logic 19
   Norihiro Kamide

Intermediate-qudit Assisted Improved Quantum Algorithm for String Matching
   with an Advanced Decomposition of Fredkin Gate 63
   Amit Saha and Om Khanna

Some Consistency Results for Many-Valued Judgment Aggregation 85
   Christian G. Fermüller and Sebastian Uhl

EDITORIAL

MARTIN LUKAC
Hiroshima City University, Japan.
malu@hiroshima-cu.ac.jp

MIKE BEHRISCH
TU Wien, Austria.
mike.behrisch@tuwien.ac.at

This year Special Issue on Multiple-Valued Logic (MVL) in Journal of Applied Logics (IfColog) appears in an exciting epoch. Quantum computer development jumped another leap., parallel processors and accelerators reached novel high speed performances, and other. Similarly this year in the MVL area we have some exciting new developments and articles. This year we have four contributions, two of them in the area of quantum-reversible circuits and two from the logic area.

The first paper is entitled *Multiple-Valued Reversible Fredkin Gates and Ensembles* explores the problem of the design of MV logic reversible circuits using a multiple-valued generalization of the binary conservative reversible Fredkin logic gates. To account for multiple multiple-valued control signals the author proposes an ensemble of multiple-valued Fredkin gates allowing to synthesize larger multiple-valued product terms. While this paper does not introduce a complete high level synthesis method, it provides a detailed account of generating ensembles of ternary as well as p-valued generalizations of Fredkin gates.

The second paper entitled *Alternative Cut-Free Sequent Calculi For S4 That Are Compatible With Paradefinite Four-Valued Logic* introduces, three new types of sequent calculus are introduced. Sequent calculus is a type of formal method applied to sequents, statements about logical relations (implications) between formulas. The The paper specifically talks about Avron's paradefinite four-valued logic as a basis for the specific extensions of the GMA4 sequent calculus. The paper in addition proves cut-elimination's well as that the S4 logic system also expresses the Gurevich logic.

The third paper is entitled *Intermediate-qudit assisted Improved quantum algorithm for string matching with an Advanced Decomposition of Fredkin gate* discusses a quantum circuit implementation for the string-matching algorithm. In particular this work proposes to use Multiple-Valued qubits (qudits) with three and four basis states respectively. Several cost optimizations are implemented including reduced query and time complexity, reduction through quantum gate optimization and error analysis.

The last paper in this year special issue on MBL is entitled *Some Consistency Results for Many-Valued Judgment Aggregation* and describes the judgment aggregation (JA) from the point of view of many-valued logics. The paper presents positive results for achieving consistent judgments in many-valued settings through Average Aggregation in Many-Valued Logics, Generalized Aggregation Rules with Systematicity and Median Aggregation and Unidimensional Alignment. The results presented in this paper confirm that using the proposed methods a consistent JA can be obtained.

We hope that this year's edition of the special issue on MVL in the JAL IfColog will be informative and exciting.

Martin Lukac
Mike Behrisch

Multiple-Valued Reversible Fredkin Gates And Ensembles

Claudio Moraga
Technical University Dortmund, 44221 Dortmund, Germany
claudio.moraga@udo.edu

Abstract

In the early 80's Edward Fredkin introduced one of the first controlled binary reversible gates, which allowed the swapping of two target signals according to the value of a control signal. Generalizations of this gate have been done in the binary domain by including several control signals, thus allowing a mixed polarity control of the gate. In the ternary domain the realization of Fredkin gates have been rarely reported. In the present work we study the Fredkin gate in the multiple-valued domain and introduce ensembles of Fredkin gates allowing several target signals. Swapping becomes a permutation, which is determined by the control signals of the gates.

1 Introduction

Power consumption and power dissipation as waste heat have become an increasingly severe problem in digital circuits realizations as the VLSI improvements are allowing millions of transistors in a chip and gigahertz switching speed.

Rolf Landauer [15] provided a formal explanation for the heating phenomenon, which has become known as Landauer's Principle:

"Any process that loses or erases 1 bit of known information increase the total entropy by at least

$$DS = k_B \times \ln(2)$$

and thus implies the eventual dissipation of at least

$$E_{diss} = T_{env} \cdot DS \geq 18 meV$$

of free energy to the environment as waste heat."

18 milli-electron-Volt seems to be a very little amount, but it refers to a bit, however, today we are working with megabytes and gigabytes. Furthermore, in the range of gigahertz.

It becomes apparent that in our classical digital systems, only inverters neither erase nor lose a bit of information, whereas AND and OR gates have several inputs and only one output, therefore losing or erasing one or more bits of information and therefore, they generate waste heat.

In 1973 Charles Bennet [3] pointed out that a computer would work "cold" if all its components were reversible. An experimental proof has been reported in [26].

Roughly 20 years after the Landauer's Principle, Edward Fredkin and Tomaso Toffoli disclosed their information preserving gates [7, 25] characterized by having the same set as domain and range and by realizing a bijection, thus having the same number of inputs and outputs and, knowing the output of a gate, the input could be correctly recovered. For this reason these gates were called **reversible**. Moreover, the work of R. Feynman [8] stimulated further research in this area. Later on, Deutsch extended the work of Toffoli to the quantum domain [5] and introduced a quantum universal gate on three qubits.

A reversible binary Fredkin gate, its symbol, (borrowed from the quantum computing community [21, 10]) and its functionality is recalled in Figure 1.

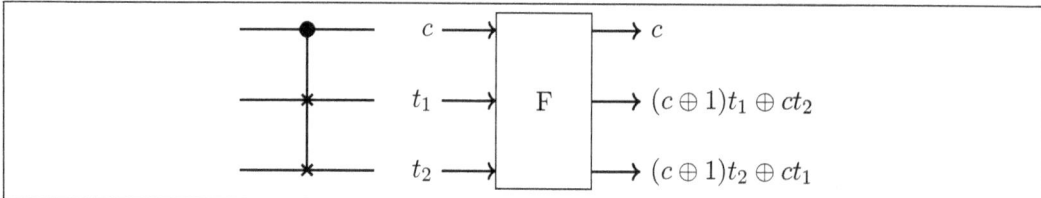

Figure 1: The binary Fredkin gate, its symbol and functionality.

From Figure 1 it may be seen that when the control signal c has the value 1, at the output side the target signals are swapped. On the other hand, when the control signal has the value 0, the gate remains inhibited and behaves as an identity. No swap takes place.

Extensions of the Fredkin gate have been studied e.g. in [4, 6] introducing mixed polarity multi- control signals, in [18] considering functional mixed polarity, in [23] with respect to quantum realizability and in [22] related to logic design. First extensions to the ternary domain have been studied in [13, 17].

The rest of the paper is organized as follows: formal aspects are presented in the next section. After that, design strategies and circuits realizing a Fredkin gate and Ensembles of Fredkin gates with two, three and four target signals are introduced.

Furthermore multiplexing ternary Fredkin ensembles is also considered. Following, quaternary Fredkin gates are discussed and p-valued generalizations are introduced. A section comprising closing remarks finishes the paper.

2 Formalisms

In the ternary domain, the basic value set is $\{0, 1, 2\}$ and the most frequently used basic reversible ternary gates realize the permutations of the value set. $\Phi = \{+1, +2, P_{12}, P_{01}, N = P_{02}\}$. The sums are modulo 3 and the indices of the permutations indicate the values which are swapped. Notice that P_{12} realizes a scaling by 2 mod 3. This set of basic gates (including the identity) is functionally complete [1]. Many authors have been using these gates for their work e.g. [19, 9, 13, 12, 14]. Φ is however not the only functionally complete set of basic gates. See e.g. [27].

Similarly, in the p-valued domain, where $p > 2$ is a natural number, the basic value set is $\{0, 1, \ldots, p-1\}$ and the set of permutations of the value set may be chosen as basic functions or gates.

Definition 1. *Let $f : (Z_p)^n \to (Z_p)^n$ be a bijection. A gate realizing such a function is a reversible gate. A reversible circuit consists of a loop-free cascade of fanout-free reversible gates.*

Definition 2. *In the circuits for quantum computing, reversibility of the gates requires a representation as unitary matrices [10, 21]. In the case of reversible circuits, the unitary requirement for the gates reduces to orthogonality.*

A complex-valued matrix M is unitary if $M \cdot (M^*)^T = I$. An integer-valued matrix M is orthogonal if $M \cdot M^T = I$, where I denotes the identity matrix of the same dimension as M.

Definition 3. *A Fredkin ensemble of basic Fredkin gates is a subcircuit in the p-valued domain, which works with one or more control signals and more than two target signals which are permuted according to given control conditions.*

Definition 4. *When basic reversible gates are controlled, they will be of the Muthukrishnan-Stroud (MS) type [20], which become activated when the control signal has the value $p - 1$, otherwise behaving as an identity. (In the last case we say that the gate is inhibited). In the circuits diagrams a gate which is activated when the control signal has the value $p - 1$, is represented with a black dot at the place where the control line meets the gate to be controlled, as shown in Figure 2.*

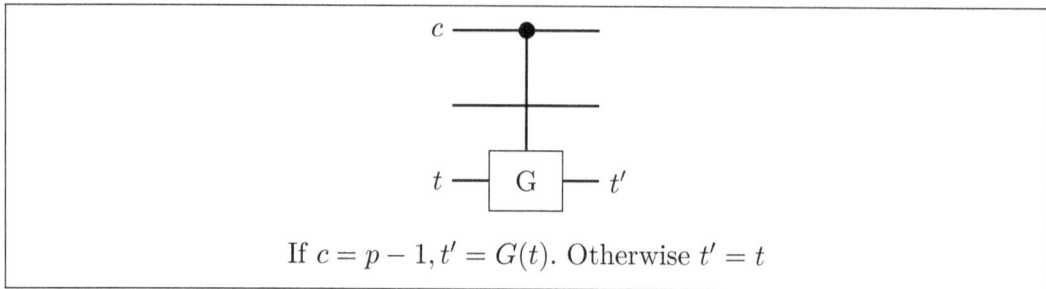

Figure 2: Representation of a p-valued reversible MS-gate controlled by c.

Definition 5. *If for some reason it is convenient that a controlled gate is activated when its control signal does not have the value $p-1$, then additional not-controlled gates are added, first to change the intended control value to the highest -(in order to activate the corresponding gate)- followed by restoring the original value.*

To avoid overloading the circuits diagrams with additional uncontrolled gates, a symbolic representation is used. In the ternary case, for example, this representation comprises grey dots when a control signal should be effective with the value 1 or white dots when a control signal should be effective with the value 0, (in analogy to the binary case [16, 24]). This is shown in Figure 3.

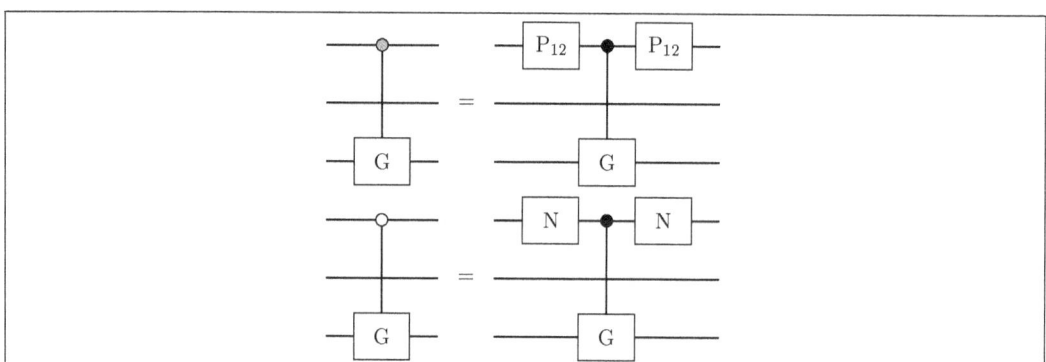

Figure 3: Symbolic representation and physical realization of ternary gates controlled by signals with value different from 2.

3 Fredkin gate and Ensembles

3.1 The basic ternary Fredkin gate.

Key components of a ternary Fredkin gate are controlled sums and subtractions modulo 3, shown in Figure 4, obtained in [17] as a controlled scaled generalization of an uncontrolled adder modulo 3 disclosed in [11].

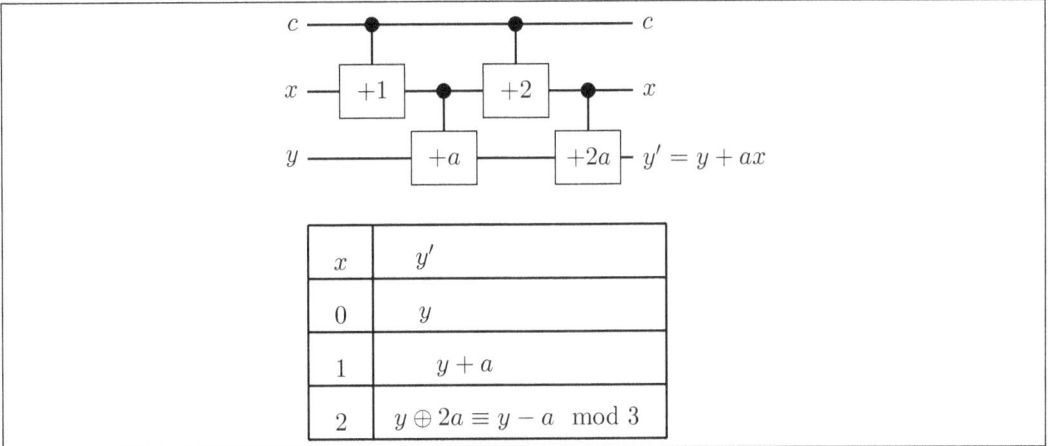

Figure 4: Controlled sums modulo 3, activated when $c = 2$ and parameterized by $a \in \{1, 2\}$

A parameter "a" $\in \{1, 2\}$ determines two functionalities. If the control signal has the value 2 and $a = 1$, the gate behaves as a (single digit) controlled modulo 3 adder and outputs the sum on the bottom line. On the other hand, if $a = 2$, the gate behaves as a (single digit) controlled subtractor modulo 3. The signal of the middle line is scaled by 2 before the modulo 3 addition and this is equivalent to a subtraction. (Recall that $-1 \equiv 2 \mod 3$). If the control signal as a value ≤ 2 the gate behaves as an identity.

A representation as "high level" gates is shown in Figure 5. Recall that $y + 2x \equiv y - x \mod 3$.

It should be noticed that "c" controls the gate and independently of the value of x, this will be added to y or subtracted from y (depending on a). In this case, the black dots on the x line do not imply that x has the value 2. Only that the value of x affects the output of the gate, either as being added or subtracted, depending on a. To avoid misunderstandings, these dots carry a diagonal.

With the high level gates it is simple to realize a ternary Freedkin gate as shown in Figure 6. These are not the only possibility. Additional realizations are available

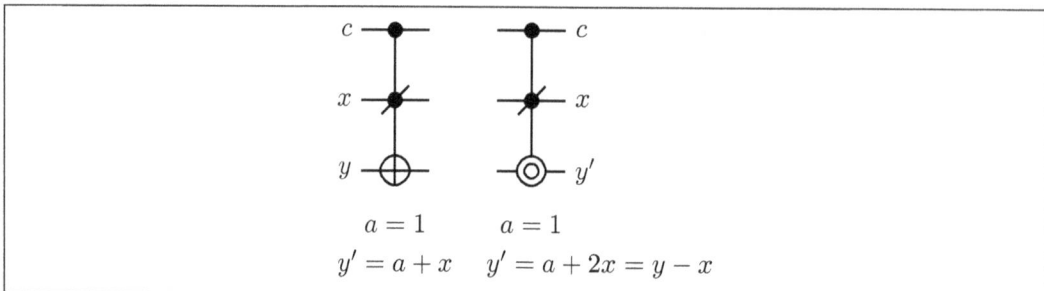

Figure 5: High level representation of the controlled sums.

in [17].

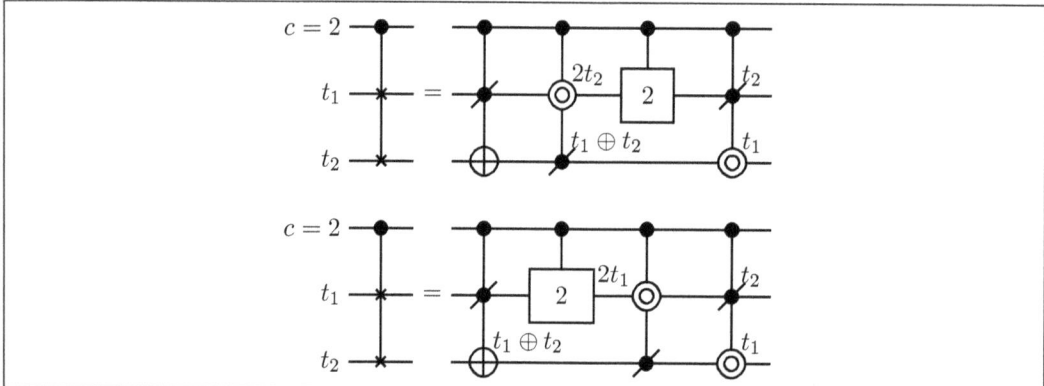

Figure 6: Possible realizations of ternary Fredkin gates.

3.2 Ternary Fredkin Ensemble.

The term Generalized Fredkin gate was used in [6] to denote multi-controlled Fredkin gates. A different generalization will be considered by which a Fredkin Ensemble of basic Fredkin gates will have more than two target lines and eventually more that one control line. Swapping will be replaced by permuting.

Case B1: A Fredkin Ensemble with one control line and three target lines to realize the permutation $(t_1, t_2, t_3) \rightarrow (t_2, t_3, t_1)$. A possible circuit is shown in Figure 7.

A fast analysis of the circuit of Figure 7 allows recognizing two (different) Fredkin gates sharing a same control line and swapping pairs of targets. Since swapping may be interpreted as a transposition, then the circuit is based on the known fact that a permutation may be realized by a sequence of appropriate transpositions.

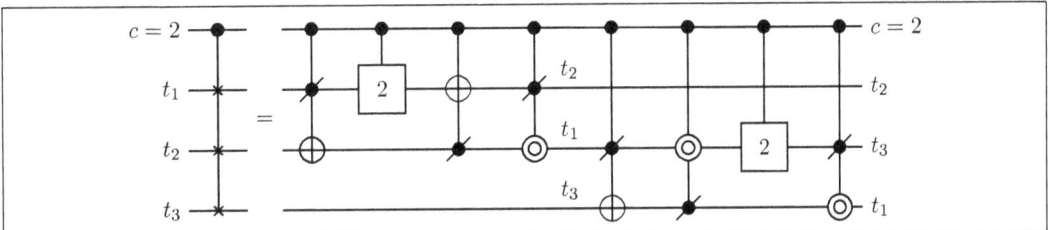

Figure 7: Fredkin based permutation $(t_1, t_2, t_3) \to (t_2, t_3, t_1)$.

It becomes apparent that Case B1 may theoretically be extended to any number of targets to be permuted. The number of needed gates and the depth of the circuit will however increase proportional to the number of targets.

On the other hand, a large number of targets may allow parallel processing of some transpositions, as discussed in the next Case.

Case B2: A Fredkin Ensemble with one control line and four target lines to realize the permutation $(t_1, t_2, t_3, t_4) \to (t_2, t_4, t_1, t_3)$. A possible circuit is shown in Figure 8.

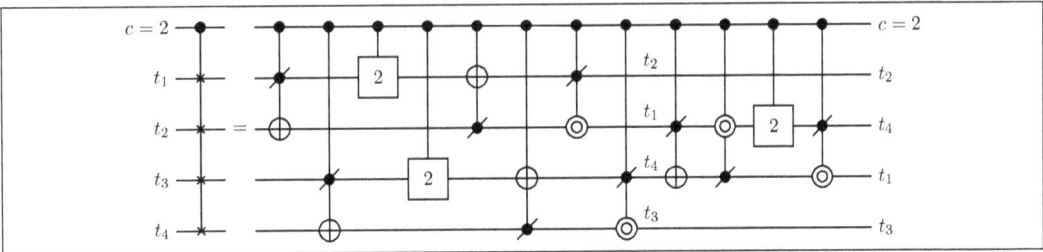

Figure 8: Fredkin based permutation $(t_1, t_2, t_3, t_4) \to (t_2, t_4, t_1, t_3)$

A careful analysis of Fig. 8 allows to recognize that in the first half there are two parallel simple Fredkin gates swapping (t_1, t_2) and (t_3, t_4), respectively, followed by a third simple (but different) Fredkin gate swapping (t_1, t_4).

Since the high level controlled sum gates comprise 4 basic ternary gates each, then the ternary Fredkin gates of Figure 6 have a "gate count" of 4 gates, but a realization complexity of 13 basic gates. The ternary Fredkin gate introduced in [13] has a realization complexity of 15 basic gates.

Similarly, the circuit of Figure 7 has a gate count of 8 and a realization complexity of 26. The circuit of Figure 8 has a gate count of 12, and a realization complexity of 39.

3.3 Multiplexing ternary Fredkin Ensembles.

As shown in section 3.2, ternary Fredkin Ensembles may be associated to permutations of target signals.

Following [1, 2], Figure 9 shows three multiplexed permutations of four target signals t_1, t_2, t_3, t_4. P_0, P_1 and P_2 denote the selected permutations, and the indices identify the value of the control signal, which will activate the corresponding permutation. Figures 9 and 10 show that depending on the value of the control

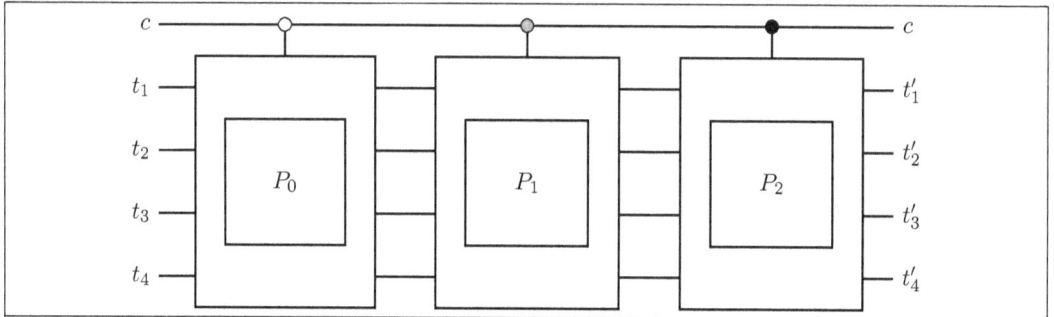

Figure 9: Multiplexed ternary Fredkin Ensemble permuting four target signals. Symbolic view.

signal one Fredkin Ensemble realizing a permutation will be activated and the other two will behave as an identity. Notice that along the control line " $+6$ " modulo 3 (i.e., 0) will be added to c. Therefore, c is recovered at the output. If $c = 0$ the first $+2$ gate locally increases its value to 2 and activates the Fredkin Ensemble realizing P_0. The other two permutations remain inhibited, since their local control signals will have the values 1 and 0, respectively. Similarly, if $c = 1$, the first $+2$ gate locally turns it to 0 and inhibits P_0. However the second $+2$ gate will increase this value to 2 and will activate the P_1 permutation. Similarly when $c = 2$ with respect to the P_2 permutation. Notice that in any case only one permutation is active and the other two are inhibited, behaving as an identity. Therefore, the active permutation receives as input t_1, t_2, t_3 and t_4 and delivers to the output, eventually through some inhibited Ensembles, the corresponding permuted vector.

Multiplexing may be in principle extended to a larger number of Fredkin Ensembles, e.g. 3^k Ensembles, if k control signals are used to activate or inhibit the Fredkin gates. Figure 11 shows a symbolic representation of a multiplexing circuit if $k = 2$. For the physical realization of the multiplexing circuit first, the two control signals must be converted to an equivalent single control signal to be applied to the Fredkin Ensemble that should be selected.

A possible realization (for the case $k = 2$) using an additional line driven by 0 is

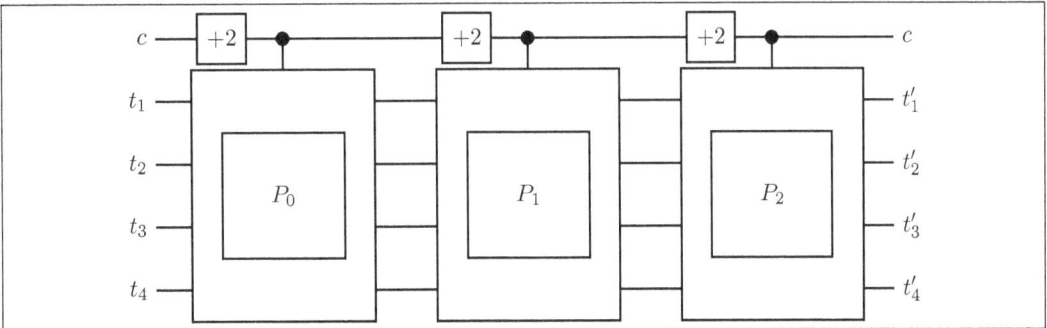

Figure 10: Physical realization of the symbolic view of Figure 9

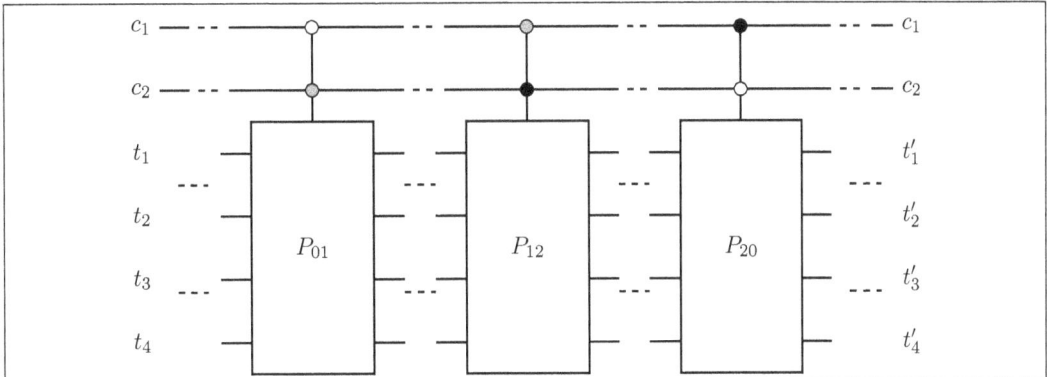

Figure 11: Symbolic representation of nine multiplexed Fredkin Ensembles permuting four target signals. (The indices of the permutations indicate the value of the control signals that activate them.)

shown in Figure 12, where the gray boxes represent +1 gates.

3.4 Quaternary Fredkin Gate and Ensemble

In the quaternary domain the value set is taken to be $\{0, 1, 2, 3\}$ and the elementary gates will realize permutations on this value set. Controlled gates will become active when $c = 3$ otherwise they will be inhibited and will behave as identities.

Following the design of a ternary Fredkin gate, a preliminary needed step is the design of controlled sum and subtraction modulo 4 gates. This is shown in Figure 13 assuming $c = 3$ and $a \in \{1, 3\}$.

As shown in the table accompanying the circuit, when $c = 3$, if $a = 1$, the target output is $y + x$, whereas if $a = 3$, the target output is $y + 3x = y - x$, since $-1 \equiv 3$ mod 4. Notice that if a were allowed to have the value 2, the circuit would not

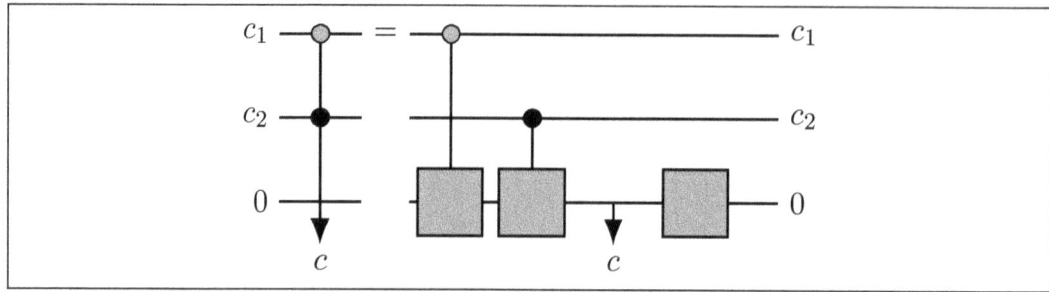

Figure 12: Simple realization of a single control signal out of two ternary control signals and an auxiliary 0-line. The grey boxes represent '+1' gates.

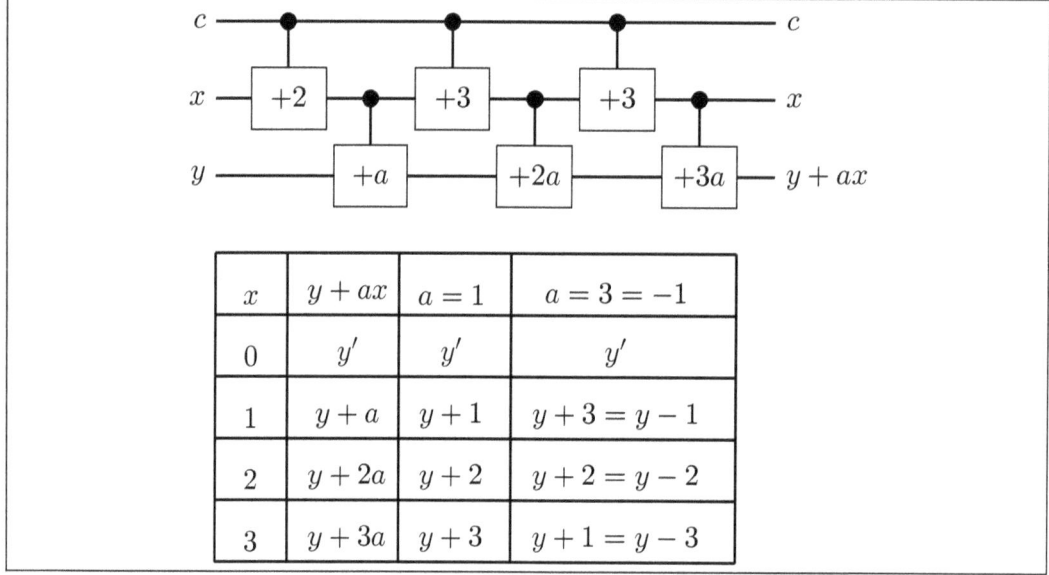

Figure 13: Controlled sums and subtractions modulo 4 parameterized by a.

distinguish between $x = 0$ and $x = 2$, since in both cases $ax \equiv 0 \mod 4$. It would also not distinguish between $x = 1$ and $x = 3$, since in both cases $ax \equiv 2 \mod 4$. Recall that 2 is a zero-divider in $(Z_4, +, \cdot)$. Therefore a is restricted to belong to $\{1, 3\}$. Under this constrain the circuit realizes $y \oplus x$ and $y \oplus 3x \equiv y - x \mod 4$.

Notice that "$y + 2x$" in the ternary case and "$y + 3x$" in the quaternary case are particular cases of "$y + (p-1)x$". Therefore, the same symbols as in the ternary case will be used for the quaternary case.

If $c = 3$:

Figure 15 shows two possible realizations of quaternary Fredkin gates. Notice

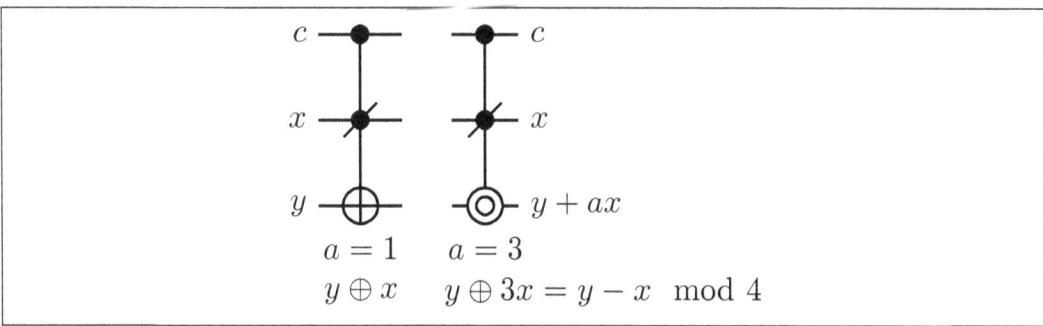

Figure 14: High level controlled sum and subtraction gates in modulo 4.

that these realizations have the same structure as the ternary circuits of Figure 6.

The controlled gate scaling by 3 is based on the permutation that swaps 1 and 3, since if $x = [0, 1, 2, 3]$ then $3x = [0, 3, 2, 1] \mod 4$. The corresponding permutation P_{13} has the following matrix representation:

$$P_{13} = Blockdiag([1], N_{(p=3)})$$

where $N_{(p=3)}$ represents the ternary negation in matrix form.

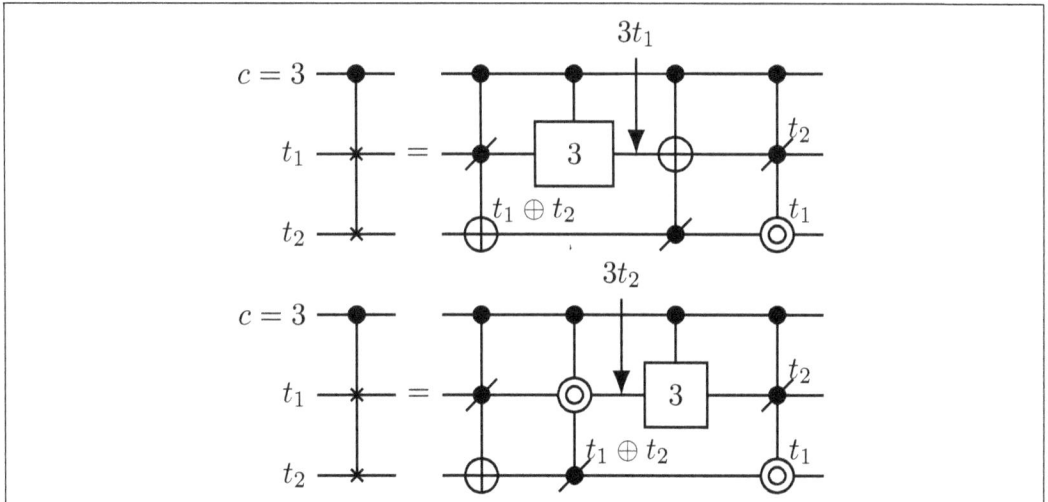

Figure 15: Two possible realizations of quaternary Fredkin gates.

3.5 p-valued Fredkin gates.

From the designs for $p = 3$ and $p = 4$ follows that for any $p > 4$ if a permutation gate $P = Blockdiag([1], N_{(p-1)})$, realizing the scaling by $(p-1)$ as well as gates for the sum and subtraction modulo p are realizable, then the following circuits in Figure 16 have the functionality of a p-valued Fredkin Gate. It is simple to prove that P is orthogonal and symmetric, therefore being self-inverse. This leads clearly to a reversible gate. Multi-target as well as multi-control Ensembles, as shown for the ternary case, are straightforward.

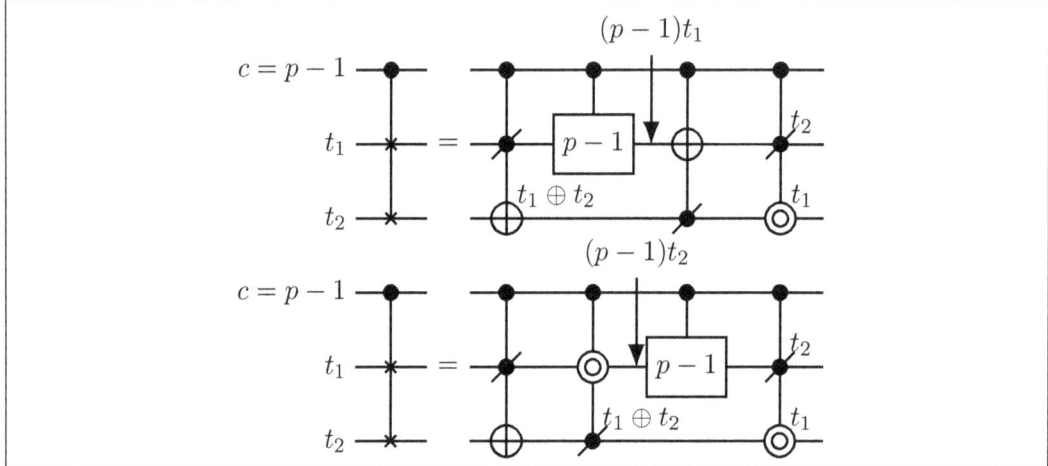

Figure 16: Two possible realizations of p-valued Fredkin gates.

It remains the design of the control-sum and control- subtractor mod p with $a \in \{1, p-1\}$.

As shown in Figures 4 and 13, for every non-0 value of x, a gate realizing "$+ax$" must be activated. In this way, at the output, $y' = y + ax$ mod p is obtained. Since a is restricted to be ± 1 mod p, a controlled adder and a controlled subtractror (both mod p), are obtained. Notice that this circuit is actually a controlled multiplexer, which for each value of x outputs ax. The general structure of this circuit is shown in Figure 17.

4 Closing Remarks

We have studied Fredkin reversible gates in the p-valued domain. We have introduced generalizations with Ensembles in terms of more than two targets, which changed swapping into controlled permutations and have considered multiplexing

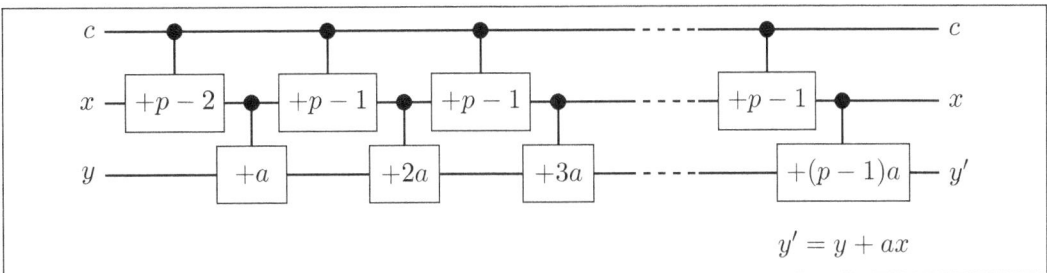

Figure 17: Structure of a p-valued controlled adder and subtractor mod p with $a \in \{1, p-1\}$.

Fredkin Ensembles in terms of one or more control signals. Since the structure of our ternary Fredkin gate is based on controlled adders and subtractors, we followed this approach to design a quaternary Fredkin gate and showed that this could be extended to any p. As required, we showed how the structure of the controlled ternary adders and subtractors could be generalized to any p.

Acknowledgment

The author appreciates the constructive criticism and suggestions for improvements given by an anonymous reviewer.

References

[1] C. Barbieri, C. Moraga, A.D. Plasencia-Lotufo and C.R. Minussi, "Comparative Analysis of the Efficiency of the Transformation Algorithm and the Cycles Decomposition Algorithm for the Synthesis of Ternary Reversible Circuits.", Journal of Multiple-Valued Logic and Soft Computing, 38, (5-6), 509-532, 2022.

[2] C. Barbieri, and C. Moraga, "Cycles-based and Transformation-based Synthesis of Ternary Reversible Circuits. Aspects of Complexity," Journal of Applied Logics, 8 (5), 1295-1309, 2021.

[3] Ch. Bennett, "Logical reversibility of computation," IBM J. of Research and Development, 17, 525-532, 1973.

[4] V. Deibuk, O. Dovhaniuk and T. Kyryliuk, "The Extended Fredkin Gates with Reconfiguration in NCT Basis," In: Hu, Z., Dychka, I., He, M. (Eds) Advances in Computer Science for Engineering and Education VI. ICCSEEA 2023. Lecture Notes on Data Engineering and Communications Technologies, Vol 181, Springer, 2023. https://doi.org/10.1007/978-3-031-36118-0_9.

[5] D. Deutsch, "Quantum computational networks," Proc. Royal Society London, A 425, 73-90, 1989.

[6] O. Dovhaniuk and V. Deibuk, "CMOS simulation of mixed-polarity Generalized Fredkin Gates". In Proc. 2022 12th International Conference on Advanced Computer Information Technologies (ACIT), IEEE Press 2022, DOI: 10.1109/ACIT54803.2022.9913119.

[7] E. Fredkin and T. Toffoli, "Conservative logic." Int. J. Theoretical Physics, 21, (3/4), 219-253, 1982.

[8] R. Feynman, "Simulating Physics with computers," Int. J. Theoretical Physics, 21, 467-488, 1982.

[9] M.M. Hawash, "Methods for Efficient Synthesis of Large Reversible Binary and Ternary Quantum Circuits and Applications of Linear Nearest Neighbor Model," Ph.D. Dissertation, Portland State University, Portland OR., 2013.

[10] M. Hirvensalo, Quantum Computing, Springer, Heidelberg, 2004.

[11] A.I. Khan et al. "Quantum realization of some ternary circuits using Muthukrishnan-Stroud Gates," in Proc. 37th Int. Symp. Multiple-valued Logic, IEEE Press, 2007.

[12] M.H.A. Kahn, M.A. Perkowski, M.R. Kahn, and P. Kerntopf, "Termary GFSOP minimization using Kronecker decision diagrams and their synthesis with quantum cascades," J. Multiple-Valued Logic and Soft Computing, 11, (5-6), 567-602, 2005

[13] M.M. Kahn "Design of Ternary Reversible/ Sequential Circuits," in Proc.8th ICECE, 2014

[14] A. B. Klimov, R. Guzman, J. C. Retamal, and C. Saavedra, "Qutrit quantum computer with trapped ions," Phys. Rev. A, 67, p. 062313, 2003.

[15] R. Landauer, "Irreversibility and heat generation in the computing process." IBM Jr. of Research and Development, 5, 183-191, 1961.

[16] D. Maslov, G. W. Dueck, D. M.. Miller, and C. Negrevergne., "Quantum circuit simplification and level compaction." IEEE Trans. CAD Integ. Circ. Syst. 27, 3, 436-444, 2008.

[17] C. Moraga, "Design of a ternary Fredkin reversible gate," Proc. 33rd IEEE Int. Conference on Micro-Electronics 125-130, IEEE Press, 2023

[18] C. Moraga and F.Z. Hadjam, "The Fredkin gate in Reversible and Quantum Environments," Facta Universitatis, Series Electronics and Energetics, 36 (2), 253-266, 2023.

[19] M. Miller and G.W. Dueck, "Search-Based Transformation Synthesis for 3-Valued Reversible Circuits," LNCS 12227, 253-266, Springer, 2023.

[20] A. Muthukrishnan and C.R. Stroud, "Multiple-valued logic gates for quantum computation," Phys. Rev. A, vol. 62, 052309, 2000.

[21] M. Nielsen, and I. Chuang, Quantum Computation and Quantum Information. Cambridge Univ. Press, UK, 2000.

[22] P.D. Picton, "Modified Fredkin gates in logic design." Microelectronics Journal, 25, 6, 437-441, 1994.

[23] J. A. Smolin and D. P. DiVincenzo. "Five two-bit quantum gates are sufficient to implement the quantum Fredkin gate". Physical Review A 53. 4. 2855-2856. DOI: 10.1103/PhysRevA.53.2855. 1996.

[24] M. Soeken and M.K. Thomsen, "White dots do matter. Rewriting Reversible Logic Circuits." RC2013, LNCS 7948, 196-208, 2013.

[25] T. Toffoli. "Reversible computing," Tech. Memo MIT/ /LCS/TM-151, MIT Lab. for Comp. Sci, 1980.

[26] C. Vieri, M.J. Ammer, M. Frank, N. Margolus and T. Knight. "A fully reversible asymptotically zero energy microprocessor," CiteSeer. 1998. DOI 10.1.1.35.474

[27] A. de Vos, B. Raa, and L. Storme, "Generating the group of reversible logic gates," Jr. Physics A : Mathematical and General, 35, 7063-7078, 2002.

Alternative cut-free sequent calculi for S4 that are compatible with paradefinite four-valued logic

Norihiro Kamide
Nagoya City University, School of Data Science,
Yamanohata 1, Mizuho-cho, Mizuho-ku, Nagoya, Aichi, 467-8501, Japan
drnkamide08@kpd.biglobe.ne.jp

Abstract

Three new Gentzen-style sequent calculi $GS4_1$, $GS4_2$, and $GS4_3$ for normal modal logic S4 are introduced. These calculi are obtained from a Gentzen-style sequent calculus GMA4 for a modal extension MA4 of Avron's self-extensional paradefinite four-valued logic by adding some special inference rules and/or some special initial sequents. The cut-elimination theorems for $GS4_1$, $GS4_2$, $GS4_3$, and GMA4 are proved. An alternative Kripke semantics for S4 is introduced and the completeness theorems with respect to this semantics are proved for $GS4_1$ and $GS4_2$. An extended Kripke semantics is introduced for an extended S4 with an auxiliary negation connective and the completeness theorem with respect to this semantics is proved for $GS4_3$. Finally, a Gödel–McKinsey–Tarski theorem for embedding Gurevich logic into S4 is proved.

1 Introduction

In this study, we introduce three new Gentzen-style sequent calculi $GS4_1$, $GS4_2$, and $GS4_3$ for normal modal logic S4 and show the cut-elimination and completeness theorems for them. Furthermore, we prove a Gödel–McKinsey–Tarski theorem for embedding Gurevich logic [21] into S4. These proposed calculi are simply obtained from a Gentzen-style sequent calculus GMA4 for a modal extension MA4 of *Avron's self-extensional paradefinite four-valued logic* SE4 [7] by adding some special inference rules and/or some special initial sequents. In this sense, the proposed calculi are regarded as compatible with SE4.

We would like to thank the anonymous referee for his or her valuable comments. This research was supported by JSPS KAKENHI Grant Number 23K10990.

The base logic SE4 [7] of the proposed calculi was originally introduced by Avron as a Tarskian consequence relation system. A Gentzen-style sequent calculus $GSE4$ for SE4 was introduced in [7]. An alternative Gentzen-style sequent calculus A4 for SE4 was also introduced in [25] as a small non-essential modification of $GSE4$. In this study, we consider a modal extension GMA4 of A4. We refer to the corresponding modal extension as *modal self-extensional paradefinite four-valued logic*, denoted as MA4 (i.e., GMA4 is regarded as a Gentzen-style sequent calculus for MA4).

The logic SE4 is known to be a specific type of *paraconsistent logics* [43] with multiple names. The paraconsistent logics of this specific type are referred to as *paradefinite logics* by Arieli and Avron [3, 4, 9], *non-alethic logics* by da Costa, and *paranormal logics* by Béziau [14]. Paradefinite logics reject both *the principle of explosion* $(\neg\alpha \wedge \alpha) \to \beta$ and *the law of excluded middle* $\neg\alpha \vee \alpha$. Through this rejection, paradefinite logics are known to be appropriate for handling inconsistent and incomplete information [3, 9].

In this study, we obtain $GS4_1$ from GMA4 by adding the *rules of explosion and excluded middle* of the form:

$$\frac{\Gamma \Rightarrow \Delta, \neg\alpha \quad \Gamma \Rightarrow \Delta, \alpha}{\Gamma \Rightarrow \Delta} \text{ (explosion)} \qquad \frac{\neg\alpha, \Gamma \Rightarrow \Delta \quad \alpha, \Gamma \Rightarrow \Delta}{\Gamma \Rightarrow \Delta} \text{ (ex-middle)}.$$

Furthermore, we obtain $GS4_2$ from GMA4 by adding the *initial sequents of explosion and excluded middle* of the form:

$$\neg\alpha, \alpha \Rightarrow \qquad \Rightarrow \neg\alpha, \alpha.$$

A similar calculus to $GS4_2$ was introduced for S4 in [28] as a modal extension of the falsification-aware normal Gentzen-style sequent calculus (for classical logic) introduced in [26]. However, the calculus introduced in [28] was not based on GMA4 (i.e., it is not compatible with SE4).

The rules (explosion) and (ex-middle) in $GS4_1$ were also introduced in [32], wherein some Gentzen-style sequent calculi for a first-order extension of SE4 were considered. The rule (ex-middle) is regarded as a classical (multi-succedent) sequent version of the intuitionistic (single-succedent) sequent version of (ex-middle) originally introduced by von Plato [45, 38]. He introduced the intuitionistic version of (ex-middle) in constructing some single-succedent Gentzen-style sequent calculi for classical logic. The intuitionistic versions of (explosion) and (ex-middle) have also recently been used in [31] in constructing a unified and modular falsification-aware single-succedent Gentzen-style framework for classical, paradefinite, paraconsistent, and paracomplete logics. Thus, using (explosion) and (ex-middle), we can obtain a unified Gentzen framework for a wide range of non-classical logics systematically.

In this study, we show the explosion- and excluded-middle-elimination theorems for Kripke's Gentzen-style sequent calculus GS4 [37] for S4. These theorems state that (explosion) and (ex-middle) are admissible in cut-free GS4. The theorems are similar to the cut-elimination theorem for Gentzen's sequent calculus LK [19] for classical logic. Using the theorems, we prove a theorem for equivalence between $GS4_1$ and GS4. Furthermore, we prove a theorem for equivalence among $GS4_1$, $GS4_2$, and GS4 and the cut- and contraposition-elimination theorems for $GS4_1$, $GS4_2$, and GMA4. Using the contraposition-elimination theorems, we show the self-extensional properties for $GS4_1$, $GS4_2$, and GMA4. We also introduce an alternative Kripke semantics for S4 and prove the completeness theorems (with respect to this semantics) for GS4, $GS4_1$, and $GS4_2$.

In this study, we also introduce $GS4_3$ for an extension of S4 with an auxiliary negation connective $\sim$ (referred to as *classical strong negation*). $GS4_3$ is constructed as a combination of $GS4_1$ and GS4. The standard classical negation connective $\neg$ and the auxiliary connective $\sim$ are equivalent in $GS4_3$ (i.e., $GS4_3 \vdash \sim\alpha \Leftrightarrow \neg\alpha$). Thus, $GS4_3$ is regarded as an alternative Gentzen-style sequent calculus for S4. Although $\sim$ in $GS4_3$ is redundant, $\sim$ is useful in proving a Gödel–McKinsey–Tarski theorem for embedding Gurevich logic into S4. Actually, $\sim$ plays a crucial role for proving this theorem (i.e., we cannot prove this theorem without using $\sim$). We prove this theorem by using a similar way as presented in [24] for proving a Gödel–McKinsey–Tarski theorem for embedding an extended intuitionistic Belnap–Dunn logic BDi with intuitionistic negation into an extended modal Belnap–Dunn logic BDm with classical negation.

The remainder of this paper is structured as follows.

In Section 2, we introduce GMA4, $GS4_1$, $GS4_2$, $GS4_3$, and GS4. We then present some basic propositions for these calculi.

In Section 3, we prove the cut-elimination theorems for GMA4, $GS4_1$, $GS4_2$, and $GS4_3$ and a theorem for equivalence among $GS4_1$, and $GS4_2$, and GS4. To prove the equivalence between $GS4_1$ and GS4, we use the explosion- and excluded-middle-elimination theorems for GS4. To prove the cut-elimination theorem for $GS4_3$, we use a theorem for syntactically embedding $GS4_3$ into GS4.

In Section 4, we prove the contraposition-elimination theorems for GMA4, $GS4_1$, $GS4_2$, $GS4_3$, and GS4. Using these contraposition-elimination theorems, we show the self-extensional properties for GMA4, $GS4_1$, $GS4_2$, $GS4_3$, and GS4.

In Section 5, we prove the completeness theorems for $GS4_1$, $GS4_2$, $GS4_3$, and GS4. First, we introduce a new alternative Kripke semantics for S4. We then present some basic propositions for this semantics. Next, we define the standard Kripke semantics for S4 and prove a theorem for equivalence between the alternative and standard semantics for S4. We then obtain the completeness theorems with respect to the alternative and standard semantics for $GS4_1$, $GS4_2$, and $GS4$. Furthermore, we introduce a new extended Kripke semantics for an extended S4 with $\sim$ and prove the completeness theorem (with respect to this extended semantics) for $GS4_3$. To prove the completeness theorem for $GS4_3$, we use some theorems for syntactically and semantically embedding $GS4_3$ into $GS4$.

In Section 6, we introduce a Gentzen-style sequent calculus GL for Gurevich logic and prove a Gödel–McKinsey–Tarski theorem for embedding GL into $GS4_3$.

In Section 7, we conclude this study and address some remarks on related works.

This paper includes some results presented in the preliminary short conference paper [29] published in the proceedings of the 53rd IEEE International Symposium on Multiple-valued Logic (ISMVL 2023). Some results on $GS4_1$ and GMA4 are presented in [29]. The results on $GS4_2$ and $GS4_3$ are newly presented in this paper.

2 Sequent calculi

We construct *formulas* of the logics discussed in this study from countably many propositional variables by the logical connectives $\land$ (conjunction), $\lor$ (disjunction), $\to$ (implication), $\neg$ (negation), $\Box$ (necessity), $\Diamond$ (possibility), and $\sim$ (classical strong negation), where $\sim$ is used in the calculus $GS4_3$ for an extended S4. We use small letters $p, q, ...$ to denote propositional variables, small Greek letters $\alpha, \beta, ...$ to denote formulas, and Greek capital letters $\Gamma, \Delta, ...$ to denote finite (possibly empty) sets of formulas. For any $\sharp \in \{\neg, \sim\}$, we use an expression $\sharp\Gamma$ to denote the set $\{\sharp\gamma \mid \gamma \in \Gamma\}$. We use the symbol $\equiv$ to denote the equality of symbols. We call an expression of the form $\Gamma \Rightarrow \Delta$ a *sequent*. We use an expression $\alpha \Leftrightarrow \beta$ to represent the abbreviation of the sequents $\alpha \Rightarrow \beta$ and $\beta \Rightarrow \alpha$. We use an expression $L \vdash S$ to denote the fact that a sequent S is provable in a sequent calculus L. We say that two sequent calculi L_1 and L_2 are *theorem-equivalent* if $\{S \mid L_1 \vdash S\} = \{S \mid L_2 \vdash S\}$. We say that a rule R of inference is *admissible* in a sequent calculus L if the following condition is satisfied: For any instance

$$\frac{S_1 \cdots S_n}{S}$$

of R, if $L \vdash S_i$ for all i, then $L \vdash S$. Furthermore, we say that R is *derivable* in L if there is a derivation from $S_1, \cdots, S_n$ to S in L. It is remarked that a rule R of inference is admissible in a sequent calculus L if and only if two sequent calculi L and $L + R$ are theorem-equivalent. Since the logics discussed in this paper are formulated as Gentzen-type sequent calculi, we will sometimes identify the logic with a Gentzen-type sequent calculus determined by it.

We introduce a Gentzen-style sequent calculus GMA4 for a modal extension MA4 of Avron's self-extensional paradefinite four-valued logic SE4.

Definition 2.1 (GMA4). *The initial sequents of* GMA4 *are of the following form for any propositional variable p:*

$$p \Rightarrow p \text{ (init1)} \qquad \neg p \Rightarrow \neg p \text{ (init2)}.$$

The structural inference rules of GMA4 *are of the form:*

$$\frac{\Gamma \Rightarrow \Delta, \alpha \quad \alpha, \Sigma \Rightarrow \Pi}{\Gamma, \Sigma \Rightarrow \Delta, \Pi} \text{ (cut)}$$

$$\frac{\Gamma \Rightarrow \Delta}{\alpha, \Gamma \Rightarrow \Delta} \text{ (we-left)} \quad \frac{\Gamma \Rightarrow \Delta}{\Gamma \Rightarrow \Delta, \alpha} \text{ (we-right)}.$$

The positive logical inference rules of GMA4 *are of the form:*

$$\frac{\alpha, \beta, \Gamma \Rightarrow \Delta}{\alpha \wedge \beta, \Gamma \Rightarrow \Delta} \text{ (}\wedge\text{left)} \quad \frac{\Gamma \Rightarrow \Delta, \alpha \quad \Gamma \Rightarrow \Delta, \beta}{\Gamma \Rightarrow \Delta, \alpha \wedge \beta} \text{ (}\wedge\text{right)}$$

$$\frac{\alpha, \Gamma \Rightarrow \Delta \quad \beta, \Gamma \Rightarrow \Delta}{\alpha \vee \beta, \Gamma \Rightarrow \Delta} \text{ (}\vee\text{left)} \quad \frac{\Gamma \Rightarrow \Delta, \alpha, \beta}{\Gamma \Rightarrow \Delta, \alpha \vee \beta} \text{ (}\vee\text{right)}$$

$$\frac{\Gamma \Rightarrow \Delta, \alpha \quad \beta, \Gamma \Rightarrow \Delta}{\alpha \rightarrow \beta, \Gamma \Rightarrow \Delta} \text{ (}\rightarrow\text{left)} \quad \frac{\alpha, \Gamma \Rightarrow \Delta, \beta}{\Gamma \Rightarrow \Delta, \alpha \rightarrow \beta} \text{ (}\rightarrow\text{right)}$$

$$\frac{\alpha, \Gamma \Rightarrow \Delta}{\Box \alpha, \Gamma \Rightarrow \Delta} \text{ (}\Box\text{left)} \quad \frac{\Box \Gamma, \neg \Diamond \Sigma \Rightarrow \Diamond \Delta, \neg \Box \Pi, \alpha}{\Box \Gamma, \neg \Diamond \Sigma \Rightarrow \Diamond \Delta, \neg \Box \Pi, \Box \alpha} \text{ (}\Box\text{right)}$$

$$\frac{\alpha, \Box \Delta, \neg \Diamond \Pi \Rightarrow \Diamond \Gamma, \neg \Box \Sigma}{\Diamond \alpha, \Box \Delta, \neg \Diamond \Pi \Rightarrow \Diamond \Gamma, \neg \Box \Sigma} \text{ (}\Diamond\text{left)} \quad \frac{\Gamma \Rightarrow \Delta, \alpha}{\Gamma \Rightarrow \Delta, \Diamond \alpha} \text{ (}\Diamond\text{right)}.$$

The negative logical inference rules of GMA4 *are of the form:*

$$\frac{\alpha, \Gamma \Rightarrow \Delta}{\neg \neg \alpha, \Gamma \Rightarrow \Delta} \text{ (}\neg\neg\text{left)} \quad \frac{\Gamma \Rightarrow \Delta, \alpha}{\Gamma \Rightarrow \Delta, \neg \neg \alpha} \text{ (}\neg\neg\text{right)}$$

$$\frac{\neg \alpha, \Gamma \Rightarrow \Delta \quad \neg \beta, \Gamma \Rightarrow \Delta}{\neg(\alpha \wedge \beta), \Gamma \Rightarrow \Delta} \text{ (}\neg\wedge\text{left)} \quad \frac{\Gamma \Rightarrow \Delta, \neg \alpha, \neg \beta}{\Gamma \Rightarrow \Delta, \neg(\alpha \wedge \beta)} \text{ (}\neg\wedge\text{right)}$$

$$\frac{\neg \alpha, \neg \beta, \Gamma \Rightarrow \Delta}{\neg(\alpha \vee \beta), \Gamma \Rightarrow \Delta} \text{ (}\neg\vee\text{left)} \quad \frac{\Gamma \Rightarrow \Delta, \neg \alpha \quad \Gamma \Rightarrow \Delta, \neg \beta}{\Gamma \Rightarrow \Delta, \neg(\alpha \vee \beta)} \text{ (}\neg\vee\text{right)}$$

$$\frac{\neg \beta, \Gamma \Rightarrow \Delta, \neg \alpha}{\neg(\alpha \rightarrow \beta), \Gamma \Rightarrow \Delta} \text{ (}\neg\rightarrow\text{left)} \quad \frac{\neg \alpha, \Gamma \Rightarrow \Delta \quad \Gamma \Rightarrow \Delta, \neg \beta}{\Gamma \Rightarrow \Delta, \neg(\alpha \rightarrow \beta)} \text{ (}\neg\rightarrow\text{right)}$$

$$\frac{\neg \alpha, \Box \Delta, \neg \Diamond \Pi \Rightarrow \Diamond \Gamma, \neg \Box \Sigma}{\neg \Box \alpha, \Box \Delta, \neg \Diamond \Pi \Rightarrow \Diamond \Gamma, \neg \Box \Sigma} \text{ (}\neg\Box\text{left)} \quad \frac{\Gamma \Rightarrow \Delta, \neg \alpha}{\Gamma \Rightarrow \Delta, \neg \Box \alpha} \text{ (}\neg\Box\text{right)}$$

$$\frac{\neg \alpha, \Gamma \Rightarrow \Delta}{\neg \Diamond \alpha, \Gamma \Rightarrow \Delta} \text{ (}\neg\Diamond\text{left)} \quad \frac{\Box \Gamma, \neg \Diamond \Sigma \Rightarrow \Diamond \Delta, \neg \Box \Pi, \neg \alpha}{\Box \Gamma, \neg \Diamond \Sigma \Rightarrow \Diamond \Delta, \neg \Box \Pi, \neg \Diamond \alpha} \text{ (}\neg\Diamond\text{right)}.$$

Remark 2.2. *The non-modal (i.e., $\{\Box, \Diamond\}$-less) fragment A4 of GMA4 was introduced in [25] as a slight non-essential modification of the original Gentzen-style sequent calculus GSE4 for Avron's self-extensional four-valued paradefinite logic SE4 [7]. The original calculus GSE4 by Avron is obtained from A4 by replacing (init1) and (init2) with the initial sequents of the form $\alpha \Rightarrow \alpha$ for any formula α.*

We have the following basic proposition for GMA4.

Proposition 2.3. *For any formula α, GMA4 − (cut) $\vdash \alpha \Rightarrow \alpha$.*

Proof. By induction on the complexity of α. We have to consider the cases for $\alpha \equiv p$, $\alpha \equiv \beta_1 \wedge \beta_2$, $\alpha \equiv \beta_1 \vee \beta_2$, $\alpha \equiv \beta_1 \rightarrow \beta_2$, $\alpha \equiv \neg\beta$, $\alpha \equiv \Box\beta$, and $\alpha \equiv \Diamond\beta$. Since the cases for $\alpha \equiv p$, $\alpha \equiv \beta_1 \wedge \beta_2$, $\alpha \equiv \beta_1 \vee \beta_2$, $\alpha \equiv \beta_1 \rightarrow \beta_2$, $\alpha \equiv \Box\beta$, and $\alpha \equiv \Diamond\beta$ are obvious, we show only the case for $\alpha \equiv \neg\beta$. To show this, we have to consider the subcases for $\beta \equiv p$, $\beta \equiv \gamma_1 \wedge \gamma_2$, $\beta \equiv \gamma_1 \vee \gamma_2$, $\beta \equiv \gamma_1 \rightarrow \gamma_2$, $\beta \equiv \neg\gamma$, $\beta \equiv \Box\gamma$, and $\beta \equiv \Diamond\gamma$. In the following, we show only the following subcases for $\beta \equiv \neg\gamma$.

1. $\gamma \equiv \gamma_1 \rightarrow \gamma_2$:

$$\dfrac{\dfrac{\dfrac{\vdots \text{ Ind.hyp.}}{\neg\gamma_1 \Rightarrow \neg\gamma_1}}{\neg\gamma_1, \neg\gamma_2 \Rightarrow \neg\gamma_1} \text{(we-left)} \quad \dfrac{\dfrac{\vdots \text{ Ind.hyp.}}{\neg\gamma_2 \Rightarrow \neg\gamma_2}}{\neg\gamma_2 \Rightarrow \neg\gamma_1, \neg\gamma_2} \text{(we-right)}}{\dfrac{\neg\gamma_2 \Rightarrow \neg(\gamma_1 \rightarrow \gamma_2), \neg\gamma_1}{\neg(\gamma_1 \rightarrow \gamma_2) \Rightarrow \neg(\gamma_1 \rightarrow \gamma_2)} \text{(}\neg\rightarrow\text{left)}} \text{(}\neg\rightarrow\text{right)}$$

2. $\gamma \equiv \Box\delta$:

$$\dfrac{\dfrac{\dfrac{\vdots \text{ Ind.hyp.}}{\neg\delta \Rightarrow \neg\delta}}{\neg\delta \Rightarrow \neg\Box\delta} \text{(}\neg\Box\text{right)}}{\neg\Box\delta \Rightarrow \neg\Box\delta} \text{(}\neg\Box\text{left)}.$$

□

Next, we introduce two new Gentzen-style sequent calculi $GS4_1$ and $GS4_2$ for normal modal logic S4.

Definition 2.4 ($GS4_1$ and $GS4_2$).

1. $GS4_1$ *is obtained from GMA4 by adding the rules of explosion and excluded middle of the form:*

$$\dfrac{\Gamma \Rightarrow \Delta, \neg\alpha \quad \Gamma \Rightarrow \Delta, \alpha}{\Gamma \Rightarrow \Delta} \text{ (explosion)} \quad \dfrac{\neg\alpha, \Gamma \Rightarrow \Delta \quad \alpha, \Gamma \Rightarrow \Delta}{\Gamma \Rightarrow \Delta} \text{ (ex-middle)}.$$

2. $GS4_2$ *is obtained from* $GMA4$ *by adding the initial sequents of explosion and excluded middle of the following form for any propositional variable p:*

$$\neg p, p \Rightarrow \text{(init3)} \qquad \Rightarrow \neg p, p \text{ (init4)}.$$

Remark 2.5. *The systems* $GS4_1$ *and* $GS4_2$ *have no standard negation inference rules* ($\neg$left) *and* ($\neg$right) *used in Gentzen's* LK *for classical logic. But, we will show in Propositions 2.9 and 2.10 that* ($\neg$left) *and* ($\neg$right) *are derivable in* $GS4_1$ *and* $GS4_2$.

We have the following basic propositions for $GS4_1$ and $GS4_2$.

Proposition 2.6. *Let* L *be* $GS4_1$ *or* $GS4_2$. *For any formula* α, $L - (\text{cut}) \vdash \alpha \Rightarrow \alpha$.

Proof. Similar to the proof of Proposition 2.3. $\square$

Proposition 2.7. *For any formula* α, *we have:*

1. $GS4_1 - (\text{cut}) \vdash \neg\alpha, \alpha \Rightarrow$,

2. $GS4_1 - (\text{cut}) \vdash \Rightarrow \neg\alpha, \alpha$.

Proof.

1. $\vdash \neg\alpha, \alpha \Rightarrow$:

$$\dfrac{\dfrac{\vdots \text{Prop. 2.6}}{\dfrac{\neg\alpha \Rightarrow \neg\alpha}{\neg\alpha, \alpha \Rightarrow \neg\alpha} \text{(we-left)}} \quad \dfrac{\vdots \text{Prop. 2.6}}{\dfrac{\alpha \Rightarrow \alpha}{\neg\alpha, \alpha \Rightarrow \alpha} \text{(we-left)}}}{\neg\alpha, \alpha \Rightarrow} \text{(explosion)}$$

2. $\vdash \Rightarrow \neg\alpha, \alpha$:

$$\dfrac{\dfrac{\vdots \text{Prop. 2.6}}{\dfrac{\neg\alpha \Rightarrow \neg\alpha}{\neg\alpha \Rightarrow \neg\alpha, \alpha} \text{(we-right)}} \quad \dfrac{\vdots \text{Prop. 2.6}}{\dfrac{\alpha \Rightarrow \alpha}{\alpha \Rightarrow \neg\alpha, \alpha} \text{(we-right)}}}{\Rightarrow \neg\alpha, \alpha} \text{(ex-middle)}.$$

$\square$

Proposition 2.8. *For any formula* α, *we have:*

1. $GS4_2 - (\text{cut}) \vdash \neg\alpha, \alpha \Rightarrow$,

2. $GS4_2 - (\text{cut}) \vdash \Rightarrow \neg\alpha, \alpha$.

Proof. We prove the statements 1 and 2 simultaneously by induction on the complexity of α. We distinguish the cases according to the form of α and show only the following cases.

1. $\alpha \equiv \beta \rightarrow \gamma$:

    (a) $\vdash \neg(\beta \rightarrow \gamma), \beta \rightarrow \gamma \Rightarrow$:

    $$\cfrac{\cfrac{\cfrac{\vdots \text{ } Ind.hyp.for \text{ } 2}{\Rightarrow \neg\beta, \beta}}{\neg\gamma \Rightarrow \neg\beta, \beta} \text{(we-left)} \quad \cfrac{\cfrac{\vdots \text{ } Ind.hyp.for \text{ } 1}{\gamma, \neg\gamma \Rightarrow}}{\gamma, \neg\gamma \Rightarrow \neg\beta} \text{(we-right)}}{\cfrac{\neg\gamma, \beta \rightarrow \gamma \Rightarrow \neg\beta}{\neg(\beta \rightarrow \gamma), \beta \rightarrow \gamma \Rightarrow} \text{($\neg\rightarrow$left)}} \text{($\rightarrow$left)}.$$

    (b) $\vdash \Rightarrow \neg(\beta \rightarrow \gamma), \beta \rightarrow \gamma$:

    $$\cfrac{\cfrac{\cfrac{\vdots \text{ } Ind.hyp.for \text{ } 1}{\neg\beta, \beta \Rightarrow}}{\neg\beta, \beta \Rightarrow \gamma} \text{(we-right)} \quad \cfrac{\cfrac{\vdots \text{ } Ind.hyp.for \text{ } 2}{\Rightarrow \neg\gamma, \gamma}}{\beta \Rightarrow \neg\gamma, \gamma} \text{(we-left)}}{\cfrac{\beta \Rightarrow \neg(\beta \rightarrow \gamma), \gamma}{\Rightarrow \neg(\beta \rightarrow \gamma), \beta \rightarrow \gamma} \text{($\rightarrow$right)}} \text{($\neg\rightarrow$right)}.$$

2. $\alpha \equiv \Box\beta$:

    (a) $\vdash \neg\Box\beta, \Box\beta \Rightarrow$:

    $$\cfrac{\cfrac{\cfrac{\vdots \text{ } Ind.hyp.for \text{ } 1}{\neg\beta, \beta \Rightarrow}}{\neg\beta, \Box\beta \Rightarrow} \text{($\Box$left)}}{\neg\Box\beta, \Box\beta \Rightarrow} \text{($\neg\Box$left)}.$$

    (b) $\vdash \Rightarrow \neg\Box\beta, \Box\beta$:

    $$\cfrac{\cfrac{\cfrac{\vdots \text{ } Ind.hyp.for \text{ } 2}{\Rightarrow \neg\beta, \beta}}{\Rightarrow \neg\Box\beta, \beta} \text{($\neg\Box$right)}}{\Rightarrow \neg\Box\beta, \Box\beta} \text{($\Box$right)}.$$

□

Proposition 2.9. *The following rules ($\neg$left) and ($\neg$right) are derivable in cut-free* $GS4_1$:

$$\cfrac{\Gamma \Rightarrow \Delta, \alpha}{\neg\alpha, \Gamma \Rightarrow \Delta} \text{($\neg$left)} \quad \cfrac{\alpha, \Gamma \Rightarrow \Delta}{\Gamma \Rightarrow \Delta, \neg\alpha} \text{($\neg$right)}.$$

Proof.

1. (¬left):

$$\cfrac{\cfrac{\vdots\ Prop.\ 2.6}{\cfrac{\neg\alpha \Rightarrow \neg\alpha}{\neg\alpha, \Gamma \Rightarrow \Delta, \neg\alpha}\ (\text{we-left}), (\text{we-right})} \quad \cfrac{\Gamma \Rightarrow \Delta, \alpha}{\neg\alpha, \Gamma \Rightarrow \Delta, \alpha}\ (\text{we-left})}{\neg\alpha, \Gamma \Rightarrow \Delta}\ (\text{explosion}).$$

2. (¬right):

$$\cfrac{\cfrac{\vdots\ Prop.\ 2.6}{\cfrac{\neg\alpha \Rightarrow \neg\alpha}{\neg\alpha, \Gamma \Rightarrow \Delta, \neg\alpha}\ (\text{we-left}), (\text{we-right})} \quad \cfrac{\alpha, \Gamma \Rightarrow \Delta}{\alpha, \Gamma \Rightarrow \Delta, \neg\alpha}\ (\text{we-right})}{\Gamma \Rightarrow \Delta, \neg\alpha}\ (\text{ex-middle}).$$

□

Proposition 2.10. *The rules (¬left) and (¬right) are derivable in* $GS4_2$ *using (cut).*

Proof.

1. (¬left):

$$\cfrac{\Gamma \Rightarrow \Delta, \alpha \quad \cfrac{\vdots\ Prop.\ 2.8}{\neg\alpha, \alpha \Rightarrow}}{\neg\alpha, \Gamma \Rightarrow \Delta}\ (\text{cut}).$$

2. (¬right):

$$\cfrac{\cfrac{\vdots\ Prop.\ 2.8}{\Rightarrow \neg\alpha, \alpha} \quad \alpha, \Gamma \Rightarrow \Delta}{\Gamma \Rightarrow \Delta, \neg\alpha}\ (\text{cut}).$$

□

Remark 2.11. *On one hand, it was shown in Propositions 2.9 and 2.10 that (¬left) and (¬right) are derivable in* $GS4_1$ *and* $GS4_2$ *using (cut). On the other hand, it can be shown that (¬left) and (¬right) are also admissible in cut-free* $GS4_1$ *and cut-free* $GS4_2$*. This fact will be obtained using Theorems 3.6 and 3.8 (the cut-elimination theorems for* $GS4_1$ *and* $GS4_2$*).*

Next, we define Kripke's Gentzen-style sequent calculus GS4 [37] for S4.

Definition 2.12 (GS4 [37]). *GS4 is obtained from* $GS4_1$ *by replacing* (init2), (explosion), (ex-middle), ($\Box$right), ($\Diamond$left), ($\neg\neg$left), ($\neg\neg$right), ($\neg\land$left), ($\neg\land$right), ($\neg\lor$left), ($\neg\lor$right), ($\neg\rightarrow$left), ($\neg\rightarrow$right), ($\neg\Box$left), ($\neg\Box$right), ($\neg\Diamond$left), ($\neg\Diamond$right), *with* ($\neg$left), ($\neg$right), *and the logical inference rules of the form:*

$$\frac{\Box\Gamma \Rightarrow \Diamond\Delta, \alpha}{\Box\Gamma \Rightarrow \Diamond\Delta, \Box\alpha} \ (\Box\text{right}^\star) \qquad \frac{\alpha, \Box\Gamma \Rightarrow \Diamond\Delta}{\Diamond\alpha, \Box\Gamma \Rightarrow \Diamond\Delta} \ (\Diamond\text{left}^\star).$$

Remark 2.13.

1. Almost the same system as GS4 was originally introduced by Kripke in [37] (p. 91) to deal with $\Box$ and $\Diamond$ simultaneously. The original system was introduced by modifying Ohnishi and Matsumoto's Gentzen-style sequent calculus introduced in [40, 41]. The system of this type has been also recently investigated and extended by Grigoriev and Petrukhin in [20].

2. Using the rules ($\Box$right*) and ($\Diamond$left*), we can show that the sequents of the form $\Box\alpha \Leftrightarrow \neg\Diamond\neg\alpha$ and $\Diamond\alpha \Leftrightarrow \neg\Box\neg\alpha$ for any formula α are provable in cut-free GS4. These sequents cannot be proved using Ohnishi and Matsumoto's system. For more information on these rules, see [37] (p. 91) and [20] (pp. 692-693).

3. We have the following fact. For any formula α, GS4 − (cut) $\vdash \alpha \Rightarrow \alpha$. This fact can be proved by induction on α. We also have the cut-elimination and completeness (w.r.t. Kripke semantics) theorems for GS4. For more information on these theorems, see [37, 20].

Next, we define $GS4_3$ as an extension of GS4. Prior to defining $GS4_3$, we extend the language as follows. We construct formulas of $GS4_3$ that are obtained from those of GS4 by adding $\sim$ (classical strong negation).

Definition 2.14 ($GS4_3$). $GS4_3$ *is obtained from* GS4 *by replacing* ($\Box$right) *and* ($\Diamond$left) *with the logical inference rules of the form:*

$$\frac{\Box\Gamma, \sim\Diamond\Sigma \Rightarrow \Diamond\Delta, \sim\Box\Pi, \alpha}{\Box\Gamma, \sim\Diamond\Sigma \Rightarrow \Diamond\Delta, \sim\Box\Pi, \Box\alpha} \ (\Box\text{right}^\sharp) \qquad \frac{\alpha, \Box\Delta, \sim\Diamond\Pi \Rightarrow \Diamond\Gamma, \sim\Box\Sigma}{\Diamond\alpha, \Box\Delta, \sim\Diamond\Pi \Rightarrow \Diamond\Gamma, \sim\Box\Sigma} \ (\Diamond\text{left}^\sharp)$$

and adding the strongly-negated initial sequents and logical inference rules of the form:

$$\sim p \Rightarrow \sim p \ (\text{init2}^\sharp) \qquad \sim p, p \Rightarrow \ (\text{init3}^\sharp) \qquad \Rightarrow \sim p, p \ (\text{init4}^\sharp)$$

$$\frac{\alpha, \Gamma \Rightarrow \Delta}{\sim\sim\alpha, \Gamma \Rightarrow \Delta} \; (\sim\sim\text{left}) \quad \frac{\Gamma \Rightarrow \Delta, \alpha}{\Gamma \Rightarrow \Delta, \sim\sim\alpha} \; (\sim\sim\text{right})$$

$$\frac{\sim\alpha, \Gamma \Rightarrow \Delta \quad \sim\beta, \Gamma \Rightarrow \Delta}{\sim(\alpha \wedge \beta), \Gamma \Rightarrow \Delta} \; (\sim\wedge\text{left}) \quad \frac{\Gamma \Rightarrow \Delta, \sim\alpha, \sim\beta}{\Gamma \Rightarrow \Delta, \sim(\alpha \wedge \beta)} \; (\sim\wedge\text{right})$$

$$\frac{\sim\alpha, \sim\beta, \Gamma \Rightarrow \Delta}{\sim(\alpha \vee \beta), \Gamma \Rightarrow \Delta} \; (\sim\vee\text{left}) \quad \frac{\Gamma \Rightarrow \Delta, \sim\alpha \quad \Gamma \Rightarrow \Delta, \sim\beta}{\Gamma \Rightarrow \Delta, \sim(\alpha \vee \beta)} \; (\sim\vee\text{right})$$

$$\frac{\sim\beta, \Gamma \Rightarrow \Delta, \sim\alpha}{\sim(\alpha \rightarrow \beta), \Gamma \Rightarrow \Delta} \; (\sim\rightarrow\text{left}) \quad \frac{\sim\alpha, \Gamma \Rightarrow \Delta \quad \Gamma \Rightarrow \Delta, \sim\beta}{\Gamma \Rightarrow \Delta, \sim(\alpha \rightarrow \beta)} \; (\sim\rightarrow\text{right})$$

$$\frac{\alpha, \Gamma \Rightarrow \Delta}{\sim\neg\alpha, \Gamma \Rightarrow \Delta} \; (\sim\neg\text{left}) \quad \frac{\Gamma \Rightarrow \Delta, \alpha}{\Gamma \Rightarrow \Delta, \sim\neg\alpha} \; (\sim\neg\text{right})$$

$$\frac{\sim\alpha, \Box\Delta, \sim\Diamond\Pi \Rightarrow \Diamond\Gamma, \sim\Box\Sigma}{\sim\Box\alpha, \Box\Delta, \sim\Diamond\Pi \Rightarrow \Diamond\Gamma, \sim\Box\Sigma} \; (\sim\Box\text{left}) \quad \frac{\Gamma \Rightarrow \Delta, \sim\alpha}{\Gamma \Rightarrow \Delta, \sim\Box\alpha} \; (\sim\Box\text{right})$$

$$\frac{\sim\alpha, \Gamma \Rightarrow \Delta}{\sim\Diamond\alpha, \Gamma \Rightarrow \Delta} \; (\sim\Diamond\text{left}) \quad \frac{\Box\Gamma, \sim\Diamond\Sigma \Rightarrow \Diamond\Delta, \sim\Box\Pi, \sim\alpha}{\Box\Gamma, \sim\Diamond\Sigma \Rightarrow \Diamond\Delta, \sim\Box\Pi, \sim\Diamond\alpha} \; (\sim\Diamond\text{right}).$$

Remark 2.15. *The strongly-negated logical inference rules except ($\sim\neg$left) and ($\sim\neg$right) in GS4$_3$ are obtained from the negative logical inference rules in GS4$_2$ by replacing $\neg$ with $\sim$. Thus, GS4$_3$ is regarded as a combination of both GS4 and GS4$_2$.*

We have the following basic propositions for GS4$_3$.

Proposition 2.16. *For any formula α, GS4$_3$ − (cut) $\vdash \alpha \Rightarrow \alpha$.*

Proof. Similar to the proof of Proposition 2.3. □

Proposition 2.17. *Let $\sharp$ be $\neg$ or $\sim$. For any formula α, we have:*

1. *GS4$_3$ − (cut) $\vdash \sharp\alpha, \alpha \Rightarrow$,*

2. *GS4$_3$ − (cut) $\vdash \Rightarrow \sharp\alpha, \alpha$.*

Proof. The proof of the cases for $\sim$ is similar to the proof of Proposition 2.7. The proof of the cases for $\neg$ is obvious. □

Proposition 2.18. *For any formula α, GS4$_3$ − (cut) $\vdash \sim\alpha \Leftrightarrow \neg\alpha$.*

Proof.

$$\frac{\vdots \; Prop. \; 2.17}{\frac{\alpha, \sim\alpha \Rightarrow}{\sim\alpha \Rightarrow \neg\alpha}} \; (\neg\text{right}) \quad \frac{\vdots \; Prop. \; 2.17}{\frac{\Rightarrow \alpha, \sim\alpha}{\neg\alpha \Rightarrow \sim\alpha}} \; (\neg\text{left}).$$

□

Remark 2.19. *Proposition 2.18 means that $\sim$ and $\neg$ are equivalent in* $GS4_3$. *Namely, $\sim$ is a redundant auxiliary connective because it can be defined by $\neg$. Thus, we can understand that* $GS4_3$ *and GS4 are essentially equivalent. Namely,* $GS4_3$ *is regarded as an alternative Gentzen-style sequent calculus for S4 with the extended auxiliary language including* $\sim$.

3 Cut-elimination

We obtain the following cut-elimination theorem for GMA4.

Theorem 3.1 (Cut-elimination for GMA4). *The rule (cut) is admissible in cut-free GMA4.*

Proof. This theorem can be proved, for example, using the standard way of Gentzen [19]. □

Remark 3.2. *The cut-elimination-theorem for Avron's original system GSE4 was proved in [7]. GSE4 is theorem-equivalent to the non-modal fragment A4 of GMA4 (i.e., A4 is a non-essential slight modification of GSE4). The cut-elimination theorem for A4 was also proved in [25].*

Next, we show the cut-elimination theorem for $GS4_1$. Prior to proving this theorem, we show a theorem for equivalence between $GS4_1$ and GS4. To prove this equivalence theorem, we have to show the following explosion-elimination and excluded-middle-elimination theorems for GS4.

Theorem 3.3 (Explosion-elimination for GS4). *The rule (explosion) is admissible in cut-free GS4.*

Proof. Suppose $GS4 - (cut) \vdash \Gamma \Rightarrow \Delta, \neg\alpha$ and $GS4 - (cut) \vdash \Gamma \Rightarrow \Delta, \alpha$. Then, we obtain $GS4 \vdash \Gamma \Rightarrow \Delta$ by using (cut):

$$\cfrac{\Gamma \Rightarrow \Delta, \neg\alpha \quad \cfrac{\vdots \quad \Gamma \Rightarrow \Delta, \alpha}{\neg\alpha, \Gamma \Rightarrow \Delta} (\neg\text{left})}{\Gamma \Rightarrow \Delta} (\text{cut}).$$

Thus, by the cut-elimination theorem for GS4, we obtain the required fact $GS4 - (cut) \vdash \Gamma \Rightarrow \Delta$. □

Theorem 3.4 (Excluded-middle-elimination for GS4). *The rule (ex-middle) is admissible in cut-free GS4.*

Proof. Suppose GS4 − (cut) ⊢ ¬α, Γ ⇒ Δ and GS4 − (cut) ⊢ α, Γ ⇒ Δ. Then, we obtain GS4 ⊢ Γ ⇒ Δ by using (cut):

$$
\cfrac{\cfrac{\vdots}{\cfrac{\neg\alpha, \Gamma \Rightarrow \Delta}{\Gamma \Rightarrow \Delta, \neg\alpha}\,(\neg\text{right})} \quad \cfrac{\vdots}{\neg\alpha, \Gamma \Rightarrow \Delta}}{\Gamma \Rightarrow \Delta}\,(\text{cut}).
$$

Thus, by the cut-elimination theorem for GS4, we obtain the required fact GS4 − (cut) ⊢ Γ ⇒ Δ. □

We then obtain the following theorem for weak equivalence between GS4$_1$ and GS4.

Theorem 3.5. *For any sequent* Γ ⇒ Δ, *we have:*

1. *if* GS4 − (cut) ⊢ Γ ⇒ Δ, *then* GS4$_1$ − (cut) ⊢ Γ ⇒ Δ,

2. *if* GS4$_1$ ⊢ Γ ⇒ Δ, *then* GS4 ⊢ Γ ⇒ Δ.

Proof.

1. We prove the statement 1 by induction on the length of the proof P of Γ ⇒ Δ in cut-free GS4. We distinguish the cases according to the last inference of P. It is sufficient to show that for the cases when the last inference of P are (¬left) and (¬right). For these cases, we obtain the required facts by Proposition 2.9.

2. We prove the statement 2 by induction on the length of the proof Q of Γ ⇒ Δ in GS4$_1$. We distinguish the cases according to the last inference of Q. We show some cases. The cases when the last inference of Q are (explosion) and (ex-middle) are obtained by Theorems 3.3 and 3.4. The cases when the last inference of Q are the negative logical inference rules can be shown using (cut). We show only the case (¬□left). Namely, the last inference of Q is of the form:

$$
\cfrac{\neg\alpha, \Box\Delta, \neg\Diamond\Pi \Rightarrow \Diamond\Gamma, \neg\Box\Sigma}{\neg\Box\alpha, \Box\Delta, \neg\Diamond\Pi \Rightarrow \Diamond\Gamma, \neg\Box\Sigma}\,(\neg\Box\text{left}).
$$

Prior to proving this case, we remark that the following converse logical inference rules for ¬ are derivable in GS4 using (cut):

$$
\cfrac{\neg\alpha, \Gamma \Rightarrow \Delta}{\Gamma \Rightarrow \Delta, \alpha}\,(\neg\text{left}^{-1}) \qquad \cfrac{\Gamma \Rightarrow \Delta, \neg\alpha}{\alpha, \Gamma \Rightarrow \Delta}\,(\neg\text{right}^{-1}).
$$

We continue to prove the case ($\neg\Box$left) as follows. By induction hypothesis, we have that GS4 $\vdash \neg\alpha, \Box\Delta, \neg\Diamond\Pi \Rightarrow \Diamond\Gamma, \neg\Box\Sigma$. Then, we obtain the required fact by:

$$
\begin{array}{c}
\vdots\ Ind.hyp. \\
\neg\alpha, \Box\Delta, \neg\Diamond\Pi \Rightarrow \Diamond\Gamma, \neg\Box\Sigma \\
\vdots\ (\neg\text{left}^{-1}), (\neg\text{right}^{-1}) \\
\dfrac{\Box\Delta, \Box\Sigma \Rightarrow \Diamond\Gamma, \Diamond\Pi, \alpha}{\Box\Delta, \Box\Sigma \Rightarrow \Diamond\Gamma, \Diamond\Pi, \Box\alpha}\ (\Box\text{right}) \\
\vdots\ (\neg\text{left}), (\neg\text{right}). \\
\neg\Box\alpha, \Box\Delta, \neg\Diamond\Pi \Rightarrow \Diamond\Gamma, \neg\Box\Sigma
\end{array}
$$

□

We obtain the following cut-elimination theorem for $GS4_1$.

Theorem 3.6 (Cut-elimination for $GS4_1$). *The rule* (cut) *is admissible in cut-free* $GS4_1$.

Proof. Suppose that $GS4_1 \vdash S$ for an arbitrary sequent S. Then, we obtain GS4 $\vdash S$ by Theorem 3.5 (2). Thus, we obtain GS4 − (cut) $\vdash S$ by the cut-elimination theorem for GS4. Thus, we obtain the required fact $GS4_1$ − (cut) $\vdash S$ by Theorem 3.5 (1). □

We also obtain the following theorem for equivalence between $GS4_1$ and GS4.

Theorem 3.7 (Equivalence between $GS4_1$ and GS4). *For any sequent* $\Gamma \Rightarrow \Delta$, *we have:*

1. GS4 − (cut) $\vdash \Gamma \Rightarrow \Delta$ *iff* $GS4_1$ − (cut) $\vdash \Gamma \Rightarrow \Delta$,

2. GS4 $\vdash \Gamma \Rightarrow \Delta$ *iff* $GS4_1 \vdash \Gamma \Rightarrow \Delta$.

Proof. By Theorems 3.5 and 3.6. □

Next, we show the following cut-elimination theorem for $GS4_2$.

Theorem 3.8 (Cut-elimination for $GS4_2$). *The rule* (cut) *is admissible in cut-free* $GS4_2$.

Proof. We can prove this theorem using the standard syntactic method by Gentzen [19]. Since the cut-elimination theorem for GMA4 holds, it is sufficient to consider the additional cases concerning the initial sequents (init3) and (init4) in the proof. These cases are obvious. □

Remark 3.9. *As will be mentioned in Remark 3.17, Theorem 3.8 can also be obtained as a corollary of Theorem 3.15 (the cut-elimination theorem for* GS4$_3$*).*

Next, we prove the following theorem.

Theorem 3.10 (Equivalence between GS4$_1$ and GS4$_2$). *For any sequent* $\Gamma \Rightarrow \Delta$, *we have:*

1. GS4$_1$ $-$ (cut) $\vdash \Gamma \Rightarrow \Delta$ *iff* GS4$_2$ $-$ (cut) $\vdash \Gamma \Rightarrow \Delta$,
2. GS4$_1$ $\vdash \Gamma \Rightarrow \Delta$ *iff* GS4$_2$ $\vdash \Gamma \Rightarrow \Delta$.

Proof. We prove only the statement 2 below. The statement 1 can be obtained from the statement 2 and Theorems 3.6 and 3.8.

($\Longrightarrow$): It is sufficient to show that (explosion) and (ex-middle) are derivable in GS4$_2$. This fact is proved as follows.

1. (explosion):

$$
\cfrac{\Gamma \Rightarrow \Delta, \alpha \qquad \cfrac{\cfrac{\Gamma \Rightarrow \Delta, \neg\alpha}{\alpha, \Gamma \Rightarrow \Delta, \neg\alpha}\,(\text{we-left}) \qquad \cfrac{\vdots\ Prop.\ 2.8 \\ \neg\alpha, \alpha \Rightarrow \\ \vdots\ (\text{we-left}), (\text{we-right}) \\ \neg\alpha, \alpha, \Gamma \Rightarrow \Delta}{}\,(\text{cut})}{\alpha, \Gamma \Rightarrow \Delta}}{\Gamma \Rightarrow \Delta}\,(\text{cut}).
$$

2. (ex-middle):

$$
\cfrac{\cfrac{\vdots\ Prop.\ 2.8 \\ \Rightarrow \alpha, \neg\alpha \\ \vdots\ (\text{we-left}), (\text{we-right}) \\ \Gamma \Rightarrow \Delta, \alpha, \neg\alpha \qquad \cfrac{\neg\alpha, \Gamma \Rightarrow \Delta}{\neg\alpha, \Gamma \Rightarrow \Delta, \alpha}\,(\text{we-right})}{\Gamma \Rightarrow \Delta, \alpha}\,(\text{cut}) \qquad \alpha, \Gamma \Rightarrow \Delta}{\Gamma \Rightarrow \Delta}\,(\text{cut}).
$$

($\Longleftarrow$): It is sufficient to show that (init3) and (init4) are provable in GS4$_1$. This fact is obtained by Proposition 2.7. $\square$

As a consequence, we obtain the following theorem.

Theorem 3.11 (Equivalence among GS4$_1$, GS4$_2$, and GS4).

1. *GS4$_1$, GS4$_2$, and GS4 are theorem-equivalent.*

2. $GS4_1 - (cut)$, $GS4_2 - (cut)$, *and* $GS4 - (cut)$ *are theorem-equivalent.*

Proof. By Theorems 3.7 and 3.10. □

Next, we show the cut-elimination theorem for $GS4_3$ by using a theorem for embedding $GS4_3$ into GS4. Prior to proving this embedding theorem, we introduce a translation from the formulas of $GS4_3$ to those of GS4.

Definition 3.12. *Let Φ be a set of propositional variables. The language (or the set of formulas) $\mathcal{L}^\sim$ of $GS4_3$ is defined using Φ, $\wedge$, $\vee$, $\rightarrow$, $\neg$, $\Box$, $\Diamond$, and $\sim$. The language (or the set of formulas) $\mathcal{L}$ of GS4 is obtained from $\mathcal{L}^\sim$ by deleting $\sim$.*

A mapping f from $\mathcal{L}^\sim$ to $\mathcal{L}$ is defined inductively by:

1. *For any $p \in \Phi$, $f(p) := p$ and $f(\sim p) := \neg p$,*

2. $f(\alpha \sharp \beta) := f(\alpha) \sharp f(\beta)$ *where* $\sharp \in \{\wedge, \vee, \rightarrow\}$,

3. $f(\sharp \alpha) := \sharp f(\alpha)$ *where* $\sharp \in \{\neg, \Box, \Diamond\}$,

4. $f(\sharp \alpha) := f(\alpha)$ *where* $\sharp \in \{\sim\sim, \sim\neg\}$,

5. $f(\sim(\alpha \wedge \beta)) := f(\sim\alpha) \vee f(\sim\beta)$,

6. $f(\sim(\alpha \vee \beta)) := f(\sim\alpha) \wedge f(\sim\beta)$,

7. $f(\sim(\alpha \rightarrow \beta)) := \neg f(\sim\alpha) \wedge f(\sim\beta)$,

8. $f(\sim\Box\alpha) := \Diamond f(\sim\alpha)$,

9. $f(\sim\Diamond\alpha) := \Box f(\sim\alpha)$.

An expression $f(\Gamma)$ denotes the result of replacing every occurrence of a formula α in Γ by an occurrence of $f(\alpha)$.

Remark 3.13. *A similar translation to the translation defined in Definition 3.12 has been used by Gurevich [21], Rautenberg [44], and Vorob'ev [46] to embed some variants of Nelson's constructive logics [1, 39] into intuitionistic logic.*

We obtain the following weak theorem for syntactically embedding $GS4_3$ into GS4.

Theorem 3.14. *Let Γ and Δ be (possibly empty) sets of formulas in $\mathcal{L}^\sim$ and f be the mapping defined in Definition 3.12.*

1. *If $GS4_3 \vdash \Gamma \Rightarrow \Delta$, then $GS4 \vdash f(\Gamma) \Rightarrow f(\Delta)$.*

2. If $\text{GS4} - (\text{cut}) \vdash f(\Gamma) \Rightarrow f(\Delta)$, then $\text{GS4}_3 - (\text{cut}) \vdash \Gamma \Rightarrow \Delta$.

Proof.

1. By induction on the length of the proofs P of $\Gamma \Rightarrow \Delta$ in GS4_3. We distinguish the cases according to the last inference of P and show only the following cases.

   (a) $(\text{init3}^\sharp)$: The last inference of P is of the form: $p, \sim p \Rightarrow$ for any $p \in \Phi$. Using $(\neg\text{left})$, we obtain $\text{GS4} \vdash p, \neg p \Rightarrow$. Thus, we obtain $\text{GS4} \vdash f(p), f(\sim p) \Rightarrow$ where $f(p)$ and $f(\sim p)$ coincide with p and $\neg p$, respectively, by the definition of f.

   (b) $(\sim\Box\text{left})$: The last inference of P is of the form:
   $$\frac{\sim\alpha, \Box\Delta, \sim\Diamond\Pi \Rightarrow \Diamond\Gamma, \sim\Box\Sigma}{\sim\Box\alpha, \Box\Delta, \sim\Diamond\Pi \Rightarrow \Diamond\Gamma, \sim\Box\Sigma} \ (\sim\Box\text{left}).$$

   By induction hypothesis, we have $\text{GS4} \vdash f(\sim\alpha), f(\Box\Delta), f(\sim\Diamond\Pi) \Rightarrow f(\Diamond\Gamma), f(\sim\Box\Sigma)$. Then, we obtain the required fact:
   $$\frac{\vdots \ Ind.hyp.}{\dfrac{f(\sim\alpha), f(\Box\Delta), f(\sim\Diamond\Pi) \Rightarrow f(\Diamond\Gamma), f(\sim\Box\Sigma)}{\Diamond f(\sim\alpha), f(\Box\Delta), f(\sim\Diamond\Pi) \Rightarrow f(\Diamond\Gamma), f(\sim\Box\Sigma)}} \ (\Diamond\text{left}^\sharp)$$
   where $\Diamond f(\sim\alpha)$ coincides with $f(\sim\Box\alpha)$ by the definition of f.

2. By induction on the length of the proofs Q of $f(\Gamma) \Rightarrow f(\Delta)$ in $\text{GS4} - (\text{cut})$. We distinguish the cases according to the last inference of Q and show only the following case.

   $(\Diamond\text{left})$: The last inference of Q is $(\Diamond\text{left})$.

   (a) The last inference of Q is of the form:
   $$\frac{f(\alpha), f(\Box\Delta), f(\sim\Diamond\Pi) \Rightarrow f(\Diamond\Gamma), f(\sim\Box\Sigma)}{f(\Diamond\alpha), f(\Box\Delta), f(\sim\Diamond\Pi) \Rightarrow f(\Diamond\Gamma), f(\sim\Box\Sigma)} \ (\Diamond\text{left})$$

   where $f(\Diamond\alpha)$ coincides with $\Diamond f(\alpha)$ by the definition of f. By induction hypothesis, we have $\text{GS4}_3 - (\text{cut}) \vdash \alpha, \Box\Delta, \sim\Diamond\Pi \Rightarrow \Diamond\Gamma, \sim\Box\Sigma$. We thus obtain the required fact:
   $$\frac{\vdots \ Ind.hyp.}{\dfrac{\alpha, \Box\Delta, \sim\Diamond\Pi \Rightarrow \Diamond\Gamma, \sim\Box\Sigma}{\Diamond\alpha, \Box\Delta, \sim\Diamond\Pi \Rightarrow \Diamond\Gamma, \sim\Box\Sigma}} \ (\Diamond\text{left}^\sharp).$$

(b) The last inference of Q is of the form:

$$\frac{f(\sim\alpha), f(\Box\Delta), f(\sim\Diamond\Pi) \Rightarrow f(\Diamond\Gamma), f(\sim\Box\Sigma)}{f(\sim\Box\alpha), f(\Box\Delta), f(\sim\Diamond\Pi) \Rightarrow f(\Diamond\Gamma), f(\sim\Box\Sigma)} \ (\Diamond\text{left})$$

where $f(\sim\Box\alpha)$ coincides with $\Diamond f(\sim\alpha)$ by the definition of f. By induction hypothesis, we have $\text{GS4}_3 - (\text{cut}) \vdash \sim\alpha, \Box\Delta, \sim\Diamond\Pi \Rightarrow \Diamond\Gamma, \sim\Box\Sigma$. We thus obtain the required fact:

$$\frac{\vdots \ Ind.hyp.}{\sim\alpha, \Box\Delta, \sim\Diamond\Pi \Rightarrow \Diamond\Gamma, \sim\Box\Sigma}{\sim\Box\alpha, \Box\Delta, \sim\Diamond\Pi \Rightarrow \Diamond\Gamma, \sim\Box\Sigma} \ (\sim\Box\text{left}).$$

□

We then obtain the following cut-elimination theorem for GS4_3.

Theorem 3.15 (Cut-elimination for GS4_3). *The rule* (cut) *is admissible in cut-free* GS4_3.

Proof. Suppose that $\text{GS4}_3 \vdash \Gamma \Rightarrow \Delta$. Then, we have $\text{GS4} \vdash f(\Gamma) \Rightarrow f(\Delta)$ by Theorem 3.14 (1), and hence $\text{GS4} - (\text{cut}) \vdash f(\Gamma) \Rightarrow f(\Delta)$ by the cut-elimination theorem for GS4. By Theorem 3.14 (2), we obtain $\text{GS4}_3 - (\text{cut}) \vdash \Gamma \Rightarrow \Delta$. □

We then obtain the following theorem for syntactically embedding GS4_3 into GS4.

Theorem 3.16 (Syntactical embedding from GS4_3 into GS4). *Let Γ and Δ be (possibly empty) sets of formulas in $\mathcal{L}^\sim$ and f be the mapping defined in Definition 3.12.*

1. $\text{GS4}_3 \vdash \Gamma \Rightarrow \Delta$ *iff* $\text{GS4} \vdash f(\Gamma) \Rightarrow f(\Delta)$.

2. $\text{GS4}_3 - (\text{cut}) \vdash \Gamma \Rightarrow \Delta$ *iff* $\text{GS4} - (\text{cut}) \vdash f(\Gamma) \Rightarrow f(\Delta)$.

Proof.

1. ($\Longrightarrow$): By Theorem 3.14 (1). ($\Longleftarrow$): Suppose $\text{GS4} \vdash f(\Gamma) \Rightarrow f(\Delta)$. Then we have $\text{GS4} - (\text{cut}) \vdash f(\Gamma) \Rightarrow f(\Delta)$ by the cut-elimination theorem for GS4. We thus obtain $\text{GS4}_3 - (\text{cut}) \vdash \Gamma \Rightarrow \Delta$ by Theorem 3.14 (2). Therefore, we have $\text{GS4}_3 \vdash \Gamma \Rightarrow \Delta$.

2. ($\Longrightarrow$): Suppose GS4$_3$ − (cut) $\vdash \Gamma \Rightarrow \Delta$. Then we have GS4$_3 \vdash \Gamma \Rightarrow \Delta$. We then obtain GS4 $\vdash f(\Gamma) \Rightarrow f(\Delta)$ by Theorem 3.14 (1). Therefore, we obtain GS4 − (cut) $\vdash f(\Gamma) \Rightarrow f(\Delta)$ by the cut-elimination theorem for GS4. ($\Longleftarrow$): By Theorem 3.14 (2).

□

Remark 3.17. *By Theorem 3.15, we can obtain the following facts: (1) GS4$_3$ is a conservative extension of GS4 and GS4$_2$ (replaced with $\sim$ and $\neg$), although $\sim$ and $\neg$ are equivalent, (2) the cut-elimination theorem for GS4$_2$ (replaced with $\sim$ and $\neg$) holds. Thus, an alternative embedding-based syntactical proof of Theorem 3.8 (the cut-elimination theorem for GS4$_2$) is obtained by Theorem 3.15.*

4 Contraposition-elimination

We prove the contraposition-elimination theorems for GMA4, GS4$_1$, GS4$_2$, GS4$_3$, and GS4. Prior to proving these theorems, we need the following proposition.

Proposition 4.1. *Let L be GMA4, GS4$_1$, GS4$_2$, GS4$_3$, or GS4. The following rules are admissible in cut-free L:*

$$\frac{\neg\neg\alpha, \Gamma \Rightarrow \Delta}{\alpha, \Gamma \Rightarrow \Delta} \ (\neg\neg\text{left}^{-1}) \quad \frac{\Gamma \Rightarrow \Delta, \neg\neg\alpha}{\Gamma \Rightarrow \Delta, \alpha} \ (\neg\neg\text{right}^{-1}).$$

Proof. We show only this proposition for GS4$_1$. Since the case for ($\neg\neg$right^{-1}) can be similarly proved as that for ($\neg\neg$left^{-1}), we show only the case for ($\neg\neg$left^{-1}) below. The proposition of this case is proved by induction on the length of the proofs P of the upper sequent $\neg\neg\alpha, \Gamma \Rightarrow \Delta$ of ($\neg\neg$left^{-1}) in cut-free GS4$_1$. We distinguish the cases according to the last inference of P and show only the following case.

(ex-middle): The last inference of P is of the form:

$$\frac{\neg\beta, \neg\neg\alpha, \Gamma \Rightarrow \Delta \quad \beta, \neg\neg\alpha, \Gamma \Rightarrow \Delta}{\neg\neg\alpha, \Gamma \Rightarrow \Delta} \ (\text{ex-middle}).$$

By induction hypotheses, we have that GS4$_1$ − (cut) $\vdash \neg\beta, \alpha, \Gamma \Rightarrow \Delta$ and GS4$_1$ − (cut) $\vdash \beta, \alpha, \Gamma \Rightarrow \Delta$. Then, we obtain the required fact by:

$$\frac{\vdots \ Ind.hyp. \quad \vdots \ Ind.hyp.}{\dfrac{\neg\beta, \alpha, \Gamma \Rightarrow \Delta \quad \beta, \alpha, \Gamma \Rightarrow \Delta}{\alpha, \Gamma \Rightarrow \Delta}} \ (\text{ex-middle}).$$

□

We then obtain the following contraposition-elimination theorems for GMA4, $GS4_1$, $GS4_2$, $GS4_3$, and GS4.

Theorem 4.2 (Contraposition-elimination for GMA4, $GS4_i$, and GS4). *Let L be GMA4, $GS4_1$, $GS4_2$, $GS4_3$, or GS4. The following rule is admissible in cut-free L:*

$$\frac{\Delta \Rightarrow \Gamma}{\neg \Gamma \Rightarrow \neg \Delta} \text{ (contraposition)}.$$

Proof. We show only this theorem for $GS4_1$ by induction on the length of the proofs P of the upper sequent $\Delta \Rightarrow \Gamma$ of (contraposition) in cut-free $GS4_1$. We show some cases.

1. (explosion): The last inference of P is of the form:

$$\frac{\Gamma \Rightarrow \Delta, \neg \alpha \quad \Gamma \Rightarrow \Delta, \alpha}{\Gamma \Rightarrow \Delta} \text{ (explosion)}.$$

By induction hypotheses, we have $GS4_1 - \text{(cut)} \vdash \neg\neg\alpha, \neg\Delta \Rightarrow \neg\Gamma$ and $GS4_1 - \text{(cut)} \vdash \neg\alpha, \neg\Delta \Rightarrow \neg\Gamma$. Then, we obtain the required fact:

$$\frac{\vdots \text{ Ind.hyp.} \qquad \vdots \text{ Ind.hyp.}}{\dfrac{\neg\neg\alpha, \neg\Delta \Rightarrow \neg\Gamma \qquad \neg\alpha, \neg\Delta \Rightarrow \neg\Gamma}{\neg\Delta \Rightarrow \neg\Gamma}} \text{ (ex-middle)}.$$

2. (ex-middle): The last inference of P is of the form:

$$\frac{\neg\alpha, \Gamma \Rightarrow \Delta \quad \alpha, \Gamma \Rightarrow \Delta}{\Gamma \Rightarrow \Delta} \text{ (ex-middle)}.$$

By induction hypotheses, we have $GS4_1 - \text{(cut)} \vdash \neg\Delta \Rightarrow \neg\Gamma, \neg\neg\alpha$ and $GS4_1 - \text{(cut)} \vdash \neg\Delta \Rightarrow \neg\Gamma, \neg\alpha$. Then, we obtain the required fact:

$$\frac{\vdots \text{ Ind.hyp.} \qquad \vdots \text{ Ind.hyp.}}{\dfrac{\neg\Delta \Rightarrow \neg\Gamma, \neg\neg\alpha \qquad \neg\Delta \Rightarrow \neg\Gamma, \neg\alpha}{\neg\Delta \Rightarrow \neg\Gamma}} \text{ (explosion)}.$$

3. ($\rightarrow$left): The last inference of P is of the form:

$$\frac{\Gamma \Rightarrow \Delta, \alpha \quad \beta, \Gamma \Rightarrow \Delta}{\alpha\rightarrow\beta, \Gamma \Rightarrow \Delta} \text{ ($\rightarrow$left)}.$$

By induction hypotheses, we have $GS4_1 - (cut) \vdash \neg\alpha, \neg\Delta \Rightarrow \neg\Gamma$ and $GS4_1 - (cut) \vdash \neg\Delta \Rightarrow \neg\Gamma, \neg\beta$. Then, we obtain the required fact:

$$\dfrac{\vdots \; Ind.hyp. \qquad \vdots \; Ind.hyp. \\ \neg\alpha, \neg\Delta \Rightarrow \neg\Gamma \qquad \neg\Delta \Rightarrow \neg\Gamma, \neg\beta}{\neg\Delta \Rightarrow \neg\Gamma, \neg(\alpha\to\beta)} \; (\neg\to\text{right}).$$

4. ($\to$right): The last inference of P is of the form:

$$\dfrac{\alpha, \Gamma \Rightarrow \Delta, \beta}{\Gamma \Rightarrow \Delta, \alpha\to\beta} \; (\to\text{right}).$$

By induction hypothesis, we have $GS4_1 - (cut) \vdash \neg\beta, \neg\Delta \Rightarrow \neg\Gamma, \neg\alpha$. Then, we obtain the required fact:

$$\dfrac{\vdots \; Ind.hyp. \\ \neg\beta, \neg\Delta \Rightarrow \neg\Gamma, \neg\alpha}{\neg(\alpha\to\beta), \neg\Delta \Rightarrow \neg\Gamma} \; (\neg\to\text{left}).$$

5. ($\neg\to$left): The last inference of P is of the form:

$$\dfrac{\neg\beta, \Gamma \Rightarrow \Delta, \neg\alpha}{\neg(\alpha\to\beta), \Gamma \Rightarrow \Delta} \; (\neg\to\text{left}).$$

By induction hypothesis, we have $GS4_1 - (cut) \vdash \neg\neg\alpha, \neg\Delta \Rightarrow \neg\Gamma, \neg\neg\beta$. Then, we obtain the required fact:

$$\dfrac{\dfrac{\dfrac{\dfrac{\vdots \; Ind.hyp. \\ \neg\neg\alpha, \neg\Delta \Rightarrow \neg\Gamma, \neg\neg\beta}{\neg\neg\alpha, \neg\Delta \Rightarrow \neg\Gamma, \beta} \; (\neg\neg\text{right}^{-1})}{\alpha, \neg\Delta \Rightarrow \neg\Gamma, \beta} \; (\neg\neg\text{left}^{-1})}{\neg\Delta \Rightarrow \neg\Gamma, \alpha\to\beta} \; (\to\text{right})}{\neg\Delta \Rightarrow \neg\Gamma, \neg\neg(\alpha\to\beta)} \; (\neg\neg\text{right})$$

where $(\neg\neg\text{left}^{-1})$ and $(\neg\neg\text{right}^{-1})$ are admissible in cut-free $GS4_1$ by Proposition 4.1.

6. ($\neg\to$right): The last inference of P is of the form:

$$\dfrac{\neg\alpha, \Gamma \Rightarrow \Delta \qquad \Gamma \Rightarrow \Delta, \neg\beta}{\Gamma \Rightarrow \Delta, \neg(\alpha\to\beta)} \; (\neg\to\text{right}).$$

By induction hypotheses, we have $GS4_1 - (cut) \vdash \neg\Delta \Rightarrow \neg\Gamma, \neg\neg\alpha$ and $GS4_1 - (cut) \vdash \neg\neg\beta, \neg\Delta \Rightarrow \neg\Gamma$. Then, we obtain the required fact:

$$\cfrac{\cfrac{\cfrac{\vdots\ Ind.hyp.}{\neg\Delta \Rightarrow \neg\Gamma, \neg\neg\alpha}}{\neg\Delta \Rightarrow \neg\Gamma, \alpha}(\neg\neg right^{-1}) \quad \cfrac{\cfrac{\vdots\ Ind.hyp.}{\neg\neg\beta, \neg\Delta \Rightarrow \neg\Gamma}}{\beta, \neg\Delta \Rightarrow \neg\Gamma}(\neg\neg left^{-1})}{\cfrac{\alpha\to\beta, \neg\Delta \Rightarrow \neg\Gamma}{\neg\neg(\alpha\to\beta), \neg\Delta \Rightarrow \neg\Gamma}(\neg\neg left)}(\to left)$$

where ($\neg\neg$left^{-1}) and ($\neg\neg$right^{-1}) are admissible in cut-free $GS4_1$ by Proposition 4.1.

7. ($\neg\Box$left): The last inference of P is of the form:

$$\cfrac{\neg\alpha, \Box\Delta, \neg\Diamond\Pi \Rightarrow \Diamond\Gamma, \neg\Box\Sigma}{\neg\Box\alpha, \Box\Delta, \neg\Diamond\Pi \Rightarrow \Diamond\Gamma, \neg\Box\Sigma}\ (\neg\Box left).$$

By induction hypothesis, we have:

$GS4_1 - (cut) \vdash \neg\Diamond\Gamma, \neg\neg\Box\Sigma \Rightarrow \neg\Box\Delta, \neg\neg\Diamond\Pi, \neg\neg\alpha.$

Then, we obtain the required fact:

$$\cfrac{\cfrac{\cfrac{\cfrac{\vdots\ Ind.hyp.}{\neg\Diamond\Gamma, \neg\neg\Box\Sigma \Rightarrow \neg\Box\Delta, \neg\neg\Diamond\Pi, \neg\neg\alpha}}{\neg\Diamond\Gamma, \neg\neg\Box\Sigma \Rightarrow \neg\Box\Delta, \neg\neg\Diamond\Pi, \alpha}(\neg\neg right^{-1})}{\neg\Diamond\Gamma, \neg\neg\Box\Sigma \Rightarrow \neg\Box\Delta, \neg\neg\Diamond\Pi, \Box\alpha}(\Box right)}{\neg\Diamond\Gamma, \neg\neg\Box\Sigma \Rightarrow \neg\Box\Delta, \neg\neg\Diamond\Pi, \neg\neg\Box\alpha}(\neg\neg right)$$

where ($\neg\neg$right^{-1}) is admissible in cut-free $GS4_1$ by Proposition 4.1.

8. ($\neg\Box$right): The last inference of P is of the form:

$$\cfrac{\Gamma \Rightarrow \Delta, \neg\alpha}{\Gamma \Rightarrow \Delta, \neg\Box\alpha}\ (\neg\Box right).$$

By induction hypothesis, we have $GS4_1 - (cut) \vdash \neg\neg\alpha, \neg\Delta \Rightarrow \neg\Gamma$. Then, we obtain the required fact:

$$\cfrac{\cfrac{\cfrac{\cfrac{\vdots\ Ind.hyp.}{\neg\neg\alpha, \neg\Delta \Rightarrow \neg\Gamma}}{\alpha, \neg\Delta \Rightarrow \neg\Gamma}(\neg\neg left^{-1})}{\Box\alpha, \neg\Delta \Rightarrow \neg\Gamma}(\Box left)}{\neg\neg\Box\alpha, \neg\Delta \Rightarrow \neg\Gamma}(\neg\neg left)$$

where ($\neg\neg$left^{-1}) is admissible in cut-free $GS4_1$ by Proposition 4.1.

☐

We also obtain the strongly-negated contraposition-elimination theorem for GS4$_3$, which states that the contraposition rule with respect to $\sim$ is admissible in cut-free GS4$_3$. Prior to proving this theorem, we need the following proposition.

Proposition 4.3. *The following rules are admissible in cut-free GS4$_3$:*

$$\frac{\sim\sim\alpha, \Gamma \Rightarrow \Delta}{\alpha, \Gamma \Rightarrow \Delta} \; (\sim\sim\text{left}^{-1}) \quad \frac{\Gamma \Rightarrow \Delta, \sim\sim\alpha}{\Gamma \Rightarrow \Delta, \alpha} \; (\sim\sim\text{right}^{-1})$$

$$\frac{\sim\neg\alpha, \Gamma \Rightarrow \Delta}{\alpha, \Gamma \Rightarrow \Delta} \; (\sim\neg\text{left}^{-1}) \quad \frac{\Gamma \Rightarrow \Delta, \sim\neg\alpha}{\Gamma \Rightarrow \Delta, \alpha} \; (\sim\neg\text{right}^{-1}).$$

Proof. Similar to the proof of Proposition 4.1. ☐

Theorem 4.4 (Strongly-negated contraposition-elimination for GS4$_3$). *The following rule is admissible in cut-free GS4$_3$:*

$$\frac{\Delta \Rightarrow \Gamma}{\sim\Gamma \Rightarrow \sim\Delta} \; (\sim\text{contraposition}).$$

Proof. Similar to the proof of Theorem 4.2. ☐

Remark 4.5.

1. *The contraposition-elimination theorem for Avron's original system GSE4 for SE4 was proved by Avron in [7]. Theorem 4.2 extends the original result to the modal case.*

2. *The contraposition-elimination theorems for the existing (non-)standard Gentzen-style sequent calculi for Nelson's paraconsistent four-valued logic N4 do not hold. Actually, the sequents $\neg(p\rightarrow q) \Leftrightarrow p \land \neg q$ for any atomic formulas p and q are provable in such a sequent calculus, but the sequents $\neg\neg(p \rightarrow q) \Leftrightarrow \neg(p \land \neg q)$ are not provable in the sequent calculus.*

3. *The contraposition-elimination theorems for other systems have been studied, for example in [34, 27, 32], wherein the contraposition-elimination theorems were proved for Gentzen-style sequent calculi for symmetric praraconsistent and extended Belnap–Dunn logics.*

We obtain the following self-extensional property for GMA4, GS4$_1$, GS4$_2$, GS4$_3$, and GS4.

Theorem 4.6 (Self-extensional property for GMA4, GS4$_i$, and GS4). *Let L be GMA4, GS4$_1$, GS4$_2$, GS4$_3$, or GS4. Let α be a subformula of a formula γ, and $\gamma^\star$ be the formula obtained from γ by replacing an occurrence of α with that of β. If $L \vdash \alpha \Leftrightarrow \beta$, then $L \vdash \gamma \Leftrightarrow \gamma^\star$.*

Proof. We use Theorems 4.2 and 4.4. We show only the theorem for GMA4 as follows. This theorem is proved by induction on γ. We show only the case when $\gamma \equiv \neg\delta$. Suppose $\vdash \alpha \Leftrightarrow \beta$. Then, by induction hypothesis, we have $L \vdash \delta \Leftrightarrow \delta^\star$. Then, we obtain the required fact $L \vdash \neg\delta \Leftrightarrow (\neg\delta)^\star$ by using the rule (contraposition), which is admissible in cut-free L by Theorem 4.2. □

Remark 4.7.

1. *The self-extensional property for SE4 was proved by Avron in [7]. Theorem 4.6 extends the original result to the modal case.*

2. *The self-extensional property for the existing (non-)standard Gentzen-style sequent calculi for Nelson's paraconsistent four-valued logic N4 does not hold. As presented previously, a counterexample is as follows. The sequents $\neg(p \rightarrow q) \Leftrightarrow p \wedge \neg q$ for some atomic formulas p and q are provable in such a sequent calculus, but the sequents $\neg\neg(p \rightarrow q) \Leftrightarrow \neg(p \wedge \neg q)$ are not provable in the sequent calculus. For more information on the self-extensional property in some variants of N4, see e.g., [47].*

3. *The self-extensional property for GS4$_1$, GS4$_2$, GS4$_3$, and GS4 is obvious because the self-extensional property for S4 is well-known.*

5 Completeness

In what follows, we use the symbol Φ to denote the set of all propositional variables (or a set of propositional variables), the symbol Φ^* to denote the set of all formulas, and the symbols $\Phi^\neg$ and $\Phi^\sim$ to denote the sets $\{\neg p \mid p \in \Phi\}$ and $\{\sim p \mid p \in \Phi\}$, respectively.

First, we introduce an alternative Kripke semantics for S4.

Definition 5.1 (Kripke S4-frame). *A structure $\langle M, R \rangle$ is called a Kripke S4-frame if*

1. *M is a non-empty set,*

2. *R is a transitive and reflexive binary relation on M.*

Definition 5.2 (Alternative S4-valuation). *An alternative S4-valuation $\models^\star$ on a Kripke S4-frame $\langle M, R \rangle$ is a mapping from $\Phi \cup \Phi^\neg$ to 2^M such that*

$$x \in \models^\star (\neg p) \text{ iff } x \notin \models^\star (p) \quad \text{(atomic classical negation condition).}$$

We will write $x \models^\star p$ and $x \models^\star \neg p$ for $x \in \models^\star (p)$ and $x \in \models^\star (\neg p)$, respectively. We will also write $x \not\models^\star p$ and $x \not\models^\star \neg p$ for $x \notin \models^\star (p)$ and $x \notin \models^\star (\neg p)$, respectively. We will also use the same notations as $x \models^\star \alpha$ and $x \not\models^\star \alpha$ for an extended alternative S4-valuation for any formula α. The alternative S4-valuation $\models^\star$ is inductively extended to the mapping from $\Phi^\star$ to 2^M by:

1. $x \models^\star \alpha \wedge \beta$ *iff* $x \models^\star \alpha$ *and* $x \models^\star \beta$,

2. $x \models^\star \alpha \vee \beta$ *iff* $x \models^\star \alpha$ *or* $x \models^\star \beta$,

3. $x \models^\star \alpha \to \beta$ *iff* $x \models^\star \alpha$ *implies* $x \models^\star \beta$,

4. $x \models^\star \Box \alpha$ *iff* $\forall y \in M \ [xRy \text{ implies } y \models^\star \alpha]$,

5. $x \models^\star \Diamond \alpha$ *iff* $\exists y \in M \ [xRy \text{ and } y \models^\star \alpha]$,

6. $x \models^\star \neg \neg \alpha$ *iff* $x \models^\star \alpha$,

7. $x \models^\star \neg(\alpha \wedge \beta)$ *iff* $x \models^\star \neg \alpha$ *or* $x \models^\star \neg \beta$,

8. $x \models^\star \neg(\alpha \vee \beta)$ *iff* $x \models^\star \neg \alpha$ *and* $x \models^\star \neg \beta$,

9. $x \models^\star \neg(\alpha \to \beta)$ *iff* $x \not\models^\star \neg \alpha$ *and* $x \models^\star \neg \beta$,

10. $x \models^\star \neg \Box \alpha$ *iff* $\exists y \in M \ [xRy \text{ and } y \models^\star \neg \alpha]$,

11. $x \models^\star \neg \Diamond \alpha$ *iff* $\forall y \in M \ [xRy \text{ implies } y \models^\star \neg \alpha]$.

Definition 5.3 (Alternative Kripke S4-model). *An alternative Kripke S4-model is a structure $\langle M, R, \models^\star \rangle$ such that*

1. $\langle M, R \rangle$ *is a Kripke S4-frame,*

2. $\models^\star$ *is an alternative S4-valuation on $\langle M, R \rangle$.*

A formula α is true *in an alternative Kripke S4-model $\langle M, R, \models^\star \rangle$ iff $x \models^\star \alpha$ for any $x \in M$, and is* a-S4-valid *in a Kripke S4-frame $\langle M, R \rangle$ iff it is true for every alternative S4-valuation $\models^\star$ on the Kripke S4-frame.*

Remark 5.4. *If the atomic classical negation condition in Definition 5.2 is deleted from the alternative Kripke semantics for* S4, *then a sound semantics for* GMA4 *is obtained. However, we have not yet obtained the completeness theorem with respect to this modified alternative semantics for* GMA4.

We have the following basic proposition for the alternative Kripke semantics.

Proposition 5.5. *For any alternative S4-valuation $\models^*$ on a Kripke S4-frame $\langle M, R \rangle$ and any formula α, $x \models^* \neg \alpha$ iff $x \not\models^* \alpha$.*

Proof. By induction on the complexity of α. We show some cases.

1. $\alpha \equiv \beta \wedge \gamma$: $x \models^* \neg(\beta \wedge \gamma)$ iff $x \models^* \neg\beta$ or $x \models^* \neg\gamma$ iff $x \not\models^* \beta$ or $x \not\models^* \gamma$ (by induction hypotheses) iff $x \not\models^* \beta \wedge \gamma$.

2. $\alpha \equiv \beta \rightarrow \gamma$: $x \models^* \neg(\beta \rightarrow \gamma)$ iff $x \not\models^* \neg\beta$ and $x \models^* \neg\gamma$ iff $x \models^* \neg\neg\beta$ and $x \not\models^* \gamma$ (by induction hypotheses) iff $x \models^* \beta$ and $x \not\models^* \gamma$ iff $x \not\models^* \beta \rightarrow \gamma$.

3. $\alpha \equiv \neg\beta$: $x \models^* \neg\neg\beta$ iff $x \models^* \beta$ iff $x \not\models^* \neg\beta$ (by induction hypothesis).

4. $\alpha \equiv \Box\beta$: $x \models^* \neg\Box\beta$ iff $\exists y \in M$ $[xRy$ and $y \models^* \neg\beta]$ iff $\exists y \in M$ $[xRy$ and $y \not\models^* \beta]$ (by induction hypothesis) iff $x \not\models^* \Box\beta$.

□

Next, we define the standard Kripke semantics for GS4.

Definition 5.6 (S4-valuation). *An S4-valuation $\models$ on a Kripke S4-frame $\langle M, R \rangle$ is a mapping from Φ to 2^M. We will write $x \models p$ and $x \not\models p$ for $x \in \models(p)$ and $x \notin \models(p)$, respectively. We will also use the same notations as $x \models \alpha$ and $x \not\models \alpha$ for an extended valuation for any formula α. The S4-valuation $\models$ is inductively extended to a mapping from Φ^* to 2^M by:*

1. $x \models \alpha \wedge \beta$ iff $x \models \alpha$ and $x \models \beta$,

2. $x \models \alpha \vee \beta$ iff $x \models \alpha$ or $x \models \beta$,

3. $x \models \alpha \rightarrow \beta$ iff $x \models \alpha$ implies $x \models \beta$,

4. $x \models \neg \alpha$ iff $x \not\models \alpha$,

5. $x \models \Box \alpha$ iff $\forall y \in M$ $[xRy$ implies $y \models \alpha]$,

6. $x \models \Diamond \alpha$ iff $\exists y \in M$ $[xRy$ and $y \models \alpha]$.

Definition 5.7 (Kripke S4-model). *A Kripke S4-model is a structure $\langle M, R, \models \rangle$ such that*

1. $\langle M, R \rangle$ *is a Kripke S4-frame,*

2. $\models$ *is an S4-valuation on $\langle M, R \rangle$.*

A formula α is true *in a Kripke S4-model $\langle M, R, \models \rangle$ iff $x \models \alpha$ for any $x \in M$, and is* S4-valid *in a Kripke S4-frame $\langle M, R \rangle$ iff it is true for every valuation $\models$ on the Kripke S4-frame.*

Remark 5.8. *The following completeness theorem holds for GS4 [37]. For any formula α, GS4 $\vdash \Rightarrow \alpha$ iff α is S4-valid.*

Next, we show the completeness theorems for $GS4_1$, $GS4_2$, and GS4. Prior to proving these theorems, we need to prove the equivalence between the standard Kripke and alternative Kripke semantics (i.e., the equivalence between the S4- and a-S4-validities). Prior to proving this equivalence theorem, we need the following lemmas.

Lemma 5.9. *For any Kripke S4-model $\langle M, R, \models \rangle$, we can construct an alternative Kripke S4-model $\langle M, R, \models^\star \rangle$ such that for any formula α and any $x \in M$, $x \models \alpha$ iff $x \models^\star \alpha$.*

Proof. Let $\langle M, R, \models \rangle$ be a Kripke S4-model. Then, we define an alternative Kripke S4-model $\langle M, R, \models^\star \rangle$ such that for any $x \in M$ and any $p \in \Phi$,

$x \models^\star p$ iff $x \models p$.

We prove this lemma by induction on the complexity of α. We show some cases.

1. $\alpha \equiv \beta \wedge \gamma$: $x \models \beta \wedge \gamma$ iff $x \models \beta$ and $x \models \gamma$ iff $x \models^\star \beta$ and $x \models^\star \gamma$ (by induction hypotheses) iff $x \models^\star \beta \wedge \gamma$.

2. $\alpha \equiv \beta \rightarrow \gamma$: $x \models \beta \rightarrow \gamma$ iff $x \models \beta$ implies $x \models \gamma$ iff $x \models^\star \beta$ implies $x \models^\star \gamma$ (by induction hypotheses) iff $x \models^\star \beta \rightarrow \gamma$.

3. $\alpha \equiv \neg \beta$: $x \models \neg \beta$ iff $x \not\models \beta$ iff $x \not\models^\star \beta$ (by induction hypothesis) iff $x \models^\star \neg \beta$ (by Proposition 5.5).

4. $\alpha \equiv \Box \beta$: $x \models \Box \beta$ iff $\forall y \in M \, [xRy$ implies $y \models \beta]$ iff $\forall y \in M \, [xRy$ implies $y \models^\star \beta]$ (by induction hypothesis) iff $x \models^\star \Box \beta$.

□

Lemma 5.10. *For any alternative Kripke S4-model $\langle M, R, \models^\star \rangle$, we can construct a Kripke S4-model $\langle M, R, \models \rangle$ such that for any formula α and any $x \in M$, $x \models^\star \alpha$ iff $x \models \alpha$.*

Proof. Similar to the proof of Lemma 5.9. □

Theorem 5.11 (Equivalence between S4- and a-S4-validities). *For any formula α, α is S4-valid iff α is a-S4-valid.*

Proof. By Lemmas 5.9 and 5.10. □

We now prove the following completeness theorems for $GS4_1$, $GS4_2$, and GS4.

Theorem 5.12 (Completeness for $GS4_1$, $GS4_2$, and GS4). *Let L be $GS4_1$, $GS4_2$, or GS4. For any formula α, we have:*

1. *$L \vdash \Rightarrow \alpha$ iff α is S4-valid,*

2. *$L \vdash \Rightarrow \alpha$ iff α is a-S4-valid.*

Proof. By the completeness theorem for GS4 and Theorems 3.11 and 5.11. □

Next, we introduce an extended Kripke semantics for an extended S4 with $\sim$.

Definition 5.13 (Extended S4-valuation). *An extended S4-valuation $\models^\bullet$ on a Kripke S4-frame $\langle M, R \rangle$ is a mapping from $\Phi \cup \Phi^\sim$ to 2^M such that*

$$x \in \models^\bullet (\sim p) \text{ iff } x \notin \models^\bullet (p) \quad (\text{atomic classical negation condition for } \sim).$$

We will use the same notations as those of Definition 5.2. The extended S4-valuation $\models^\bullet$ is inductively extended to the mapping from Φ^* to 2^M by:

1. *$x \models^\bullet \alpha \wedge \beta$ iff $x \models^\bullet \alpha$ and $x \models^\bullet \beta$,*

2. *$x \models^\bullet \alpha \vee \beta$ iff $x \models^\bullet \alpha$ or $x \models^\bullet \beta$,*

3. *$x \models^\bullet \alpha \rightarrow \beta$ iff $x \models^\bullet \alpha$ implies $x \models^\bullet \beta$,*

4. *$x \models^\bullet \neg \alpha$ iff $x \not\models^\bullet \alpha$,*

5. *$x \models^\bullet \Box \alpha$ iff $\forall y \in M \, [xRy \text{ implies } y \models^\bullet \alpha]$,*

6. *$x \models^\bullet \Diamond \alpha$ iff $\exists y \in M \, [xRy \text{ and } y \models^\bullet \alpha]$,*

7. *$x \models^\bullet \sim\sim \alpha$ iff $x \models^\bullet \alpha$,*

8. $x \models^\bullet \sim(\alpha \wedge \beta)$ iff $x \models^\bullet \sim\alpha$ or $x \models^\bullet \sim\beta$,

9. $x \models^\bullet \sim(\alpha \vee \beta)$ iff $x \models^\bullet \sim\alpha$ and $x \models^\bullet \sim\beta$,

10. $x \models^\bullet \sim(\alpha \rightarrow \beta)$ iff $x \not\models^\bullet \sim\alpha$ and $x \models^\bullet \sim\beta$,

11. $x \models^\bullet \sim\neg\alpha$ iff $x \models^\bullet \alpha$,

12. $x \models^\bullet \sim\Box\alpha$ iff $\exists y \in M\ [xRy$ and $y \models^\bullet \sim\alpha]$,

13. $x \models^\bullet \sim\Diamond\alpha$ iff $\forall y \in M\ [xRy$ implies $y \models^\bullet \sim\alpha]$.

Definition 5.14 (Extended Kripke S4-model). *An extended Kripke S4-model is a structure $\langle M, R, \models^\bullet \rangle$ such that*

1. *$\langle M, R \rangle$ is a Kripke S4-frame,*

2. *$\models^\bullet$ is an extended S4-valuation on $\langle M, R \rangle$.*

A formula α is true in an extended Kripke S4-model $\langle M, R, \models^\bullet \rangle$ iff $x \models^\bullet \alpha$ for any $x \in M$, and is e-S4-valid in a Kripke S4-frame $\langle M, R \rangle$ iff it is true for every extended S4-valuation $\models^\bullet$ on the Kripke S4-frame.

We have the following basic proposition for the extended Kripke semantics.

Proposition 5.15. *For any extended S4-valuation $\models^\bullet$ on a Kripke S4-frame $\langle M, R \rangle$ and any formula α, $x \models^\bullet \sim\alpha$ iff $x \not\models^\bullet \alpha$.*

Proof. Similar to the proof of Proposition 5.5. □

Next, we prove the completeness theorem for $GS4_3$ with respect to the extended Kripke semantics. Prior to proving this theorem, we show a theorem for semantically embedding $GS4_3$ into GS4.

Lemma 5.16. *Let f be the mapping defined in Definition 3.12. For any extended Kripke model $\langle M, R, \models^\bullet \rangle$, we can construct a Kripke model $\langle M, R, \models \rangle$ such that for any formula α and any $x \in M$, $x \models^\bullet \alpha$ iff $x \models f(\alpha)$.*

Proof. Let Φ be a set of propositional variables and Φ' be the set $\{p' \mid p \in \Phi\}$ of propositional variables. Suppose that $\langle M, R, \models^\bullet \rangle$ is an extended Kripke model where $\models^\bullet$ is a mapping from $\Phi \cup \Phi^\sim$ to 2^M. Suppose that $\langle M, R, \models \rangle$ is a Kripke model where $\models$ is a mapping from $\Phi \cup \Phi'$ to 2^M such that for any $x \in M$ and any $p \in \Phi$,

1. $x \models^\bullet p$ iff $x \models p$,

2. $x \models^\bullet {\sim}p$ iff $x \models p'$.

Then, the lemma is proved by induction on the complexity of α. We show some cases.

1. $\alpha \equiv {\sim}p$ where p is a propositional variable: $x \models^\bullet {\sim}p$ iff $x \models p'$ (by the assumption) iff $x \models f({\sim}p)$ (by the definition of f).

2. $\alpha \equiv \Box\beta$: $x \models^\bullet \Box\beta$ iff $\forall y \in M[xRy$ implies $y \models^\bullet \beta]$ iff $\forall y \in M[xRy$ implies $y \models f(\beta)]$ (by induction hypothesis) iff $x \models \Box f(\beta)$ iff $x \models f(\Box\beta)$ (by the definition of f).

3. $\alpha \equiv {\sim}\neg\beta$: $x \models^\bullet {\sim}\neg\beta$ iff $x \not\models^\bullet ({\sim}\beta)$ iff $x \not\models f({\sim}\beta)$ (by induction hypothesis) iff $x \models \neg f({\sim}\beta)$ iff $x \models f({\sim}\neg\beta)$ (by the definition of f).

4. $\alpha \equiv {\sim}(\beta{\rightarrow}\gamma)$: $x \models^\bullet {\sim}(\beta{\rightarrow}\gamma)$ iff $x \not\models^\bullet {\sim}\beta$ and $x \models^\bullet {\sim}\gamma$ iff $x \not\models f({\sim}\beta)$ and $x \models f({\sim}\gamma)$ (by induction hypotheses) iff $x \models \neg f({\sim}\beta)$ and $x \models f({\sim}\gamma)$ iff $x \models \neg f({\sim}\beta) \wedge f({\sim}\beta)$ iff $x \models f({\sim}(\beta{\rightarrow}\gamma))$ (by the definition of f).

5. $\alpha \equiv {\sim}\Box\beta$: $x \models^\bullet {\sim}\Box\beta$ iff $\exists y \in M[xRy$ and $y \models^\bullet {\sim}\beta]$ iff $\exists y \in M[xRy$ and $y \models f({\sim}\beta)]$ (by induction hypothesis) iff $x \models \Diamond f({\sim}\beta)$ iff $x \models f({\sim}\Box\beta)$ (by the definition of f).

□

Lemma 5.17. *Let f be the mapping defined in Definition 3.12. For any Kripke model $\langle M, R, \models\rangle$, we can construct an extended Kripke model $\langle M, R, \models^\bullet\rangle$ such that for any formula α and any $x \in M$, $x \models f(\alpha)$ iff $x \models^\bullet \alpha$.*

Proof. Similar to the proof of Lemma 5.16. □

Theorem 5.18 (Semantical embedding from GS4$_3$ into GS4). *Let f be the mapping defined in Definition 3.12. For any formula α, α is e-S4-valid iff $f(\alpha)$ is S4-valid.*

Proof. By Lemmas 5.16 and 5.17. □

Theorem 5.19 (Completeness for GS4$_3$). *For any formula α, GS4$_3 \vdash \Rightarrow \alpha$ iff α is e-S4-valid.*

Proof. GS4$_3 \vdash \Rightarrow \alpha$ iff GS4 $\vdash \Rightarrow f(\alpha)$ (by Theorem 3.16) iff $f(\alpha)$ is S4-valid (by the completeness theorem for GS4) iff α is e-S4-valid (by Theorem 5.18). □

6 Gödel–McKinsey–Tarski embedding

First, we introduce a Gentzen-style sequent calculus GL for Gurevich logic. We construct *formulas* of Gurevich logic from countably many propositional variables by the logical connectives $\land, \lor, \to, \neg$, and $\sim$ (strong negation). We call an expression of the form $\Gamma \Rightarrow \gamma$ (γ is a formula or the empty set) an *intuitionistic sequent* (*sequent* for short). We use the same rule names of GS4 as those for some rules of GL.

Definition 6.1 (GL). *In the following definition, we use the symbol γ to represent an arbitrary formula or the empty set.*

The initial sequents of GL *are of the following form for any propositional variable p:*

$$p \Rightarrow p \text{ (init1)} \qquad \sim p \Rightarrow \sim p \text{ (init2}^\sharp) \qquad p, \sim p \Rightarrow \text{ (init3}^\sharp).$$

The structural inference rules of GL *are of the form:*

$$\frac{\Gamma \Rightarrow \alpha \quad \alpha, \Sigma \Rightarrow \gamma}{\Gamma, \Sigma \Rightarrow \gamma} \text{ (cut)} \quad \frac{\Gamma \Rightarrow \gamma}{\alpha, \Gamma \Rightarrow \gamma} \text{ (we-left)} \quad \frac{\Gamma \Rightarrow}{\Gamma \Rightarrow \alpha} \text{ (we-right)}.$$

The normal logical inference rules of GL *are of the form:*

$$\frac{\alpha, \Gamma \Rightarrow \gamma}{\alpha \land \beta, \Gamma \Rightarrow \gamma} \text{ ($\land$left1)} \quad \frac{\beta, \Gamma \Rightarrow \gamma}{\alpha \land \beta, \Gamma \Rightarrow \gamma} \text{ ($\land$left2)}$$

$$\frac{\Gamma \Rightarrow \alpha \quad \Gamma \Rightarrow \beta}{\Gamma \Rightarrow \alpha \land \beta} \text{ ($\land$right)} \quad \frac{\alpha, \Gamma \Rightarrow \gamma \quad \beta, \Gamma \Rightarrow \gamma}{\alpha \lor \beta, \Gamma \Rightarrow \gamma} \text{ ($\lor$left)}$$

$$\frac{\Gamma \Rightarrow \alpha}{\Gamma \Rightarrow \alpha \lor \beta} \text{ ($\lor$right1)} \quad \frac{\Gamma \Rightarrow \beta}{\Gamma \Rightarrow \alpha \lor \beta} \text{ ($\lor$right2)}$$

$$\frac{\Gamma \Rightarrow \alpha \quad \beta, \Sigma \Rightarrow \gamma}{\alpha \to \beta, \Gamma, \Sigma \Rightarrow \gamma} \text{ ($\to$left)} \quad \frac{\alpha, \Gamma \Rightarrow \beta}{\Gamma \Rightarrow \alpha \to \beta} \text{ ($\to$right)}$$

$$\frac{\Gamma \Rightarrow \alpha}{\neg \alpha, \Gamma \Rightarrow} \text{ ($\neg$left)} \quad \frac{\alpha, \Gamma \Rightarrow}{\Gamma \Rightarrow \neg \alpha} \text{ ($\neg$right)}.$$

The strongly-negated logical inference rules of GL *are of the form:*

$$\frac{\alpha, \Gamma \Rightarrow \gamma}{\sim\sim\alpha, \Gamma \Rightarrow \gamma} \text{ ($\sim\sim$left)} \quad \frac{\Gamma \Rightarrow \alpha}{\Gamma \Rightarrow \sim\sim\alpha} \text{ ($\sim\sim$right)}$$

$$\frac{\sim\alpha, \Gamma \Rightarrow \gamma \quad \sim\beta, \Gamma \Rightarrow \gamma}{\sim(\alpha \land \beta), \Gamma \Rightarrow \gamma} \text{ ($\sim\land$left)}$$

$$\frac{\Gamma \Rightarrow \sim\alpha}{\Gamma \Rightarrow \sim(\alpha \land \beta)} \text{ ($\sim\land$right1)} \quad \frac{\Gamma \Rightarrow \sim\beta}{\Gamma \Rightarrow \sim(\alpha \land \beta)} \text{ ($\sim\land$right2)}$$

$$\frac{{\sim}\alpha, \Gamma \Rightarrow \gamma}{{\sim}(\alpha \vee \beta), \Gamma \Rightarrow \gamma} \ ({\sim}\vee\text{left1}) \quad \frac{{\sim}\beta, \Gamma \Rightarrow \gamma}{{\sim}(\alpha \vee \beta), \Gamma \Rightarrow \gamma} \ ({\sim}\vee\text{left2})$$

$$\frac{\Gamma \Rightarrow {\sim}\alpha \quad \Gamma \Rightarrow {\sim}\beta}{\Gamma \Rightarrow {\sim}(\alpha \vee \beta)} \ ({\sim}\vee\text{right})$$

$$\frac{\alpha, \Gamma \Rightarrow \gamma}{{\sim}(\alpha \to \beta), \Gamma \Rightarrow \gamma} \ ({\sim}{\to}\text{left1}) \quad \frac{{\sim}\beta, \Gamma \Rightarrow \gamma}{{\sim}(\alpha \to \beta), \Gamma \Rightarrow \gamma} \ ({\sim}{\to}\text{left2})$$

$$\frac{\Gamma \Rightarrow \alpha \quad \Gamma \Rightarrow {\sim}\beta}{\Gamma \Rightarrow {\sim}(\alpha \to \beta)} \ ({\sim}{\to}\text{right})$$

$$\frac{\alpha, \Gamma \Rightarrow \delta}{{\sim}{\neg}\alpha, \Gamma \Rightarrow \delta} \ ({\sim}{\neg}\text{left}) \quad \frac{\Gamma \Rightarrow \alpha}{\Gamma \Rightarrow {\sim}{\neg}\alpha} \ ({\sim}{\neg}\text{right}).$$

Remark 6.2.

1. A Gentzen-style sequent calculus $\overline{G}$ originally introduced by Gurevich [21] used the following logical inference rule instead of $(\text{init3}^\sharp)$:

$$\frac{\Gamma \Rightarrow \alpha}{{\sim}\alpha, \Gamma \Rightarrow} \ ({\sim}\text{left}).$$

The cut-elimination and completeness (w.r.t. Kripke semantics) theorems for $\overline{G}$ were proved in [21]. GL and (the propositional fragment of) $\overline{G}$ are theorem-equivalent.

2. We have the following fact. For any formula α, GL − (cut) $\vdash \alpha \Rightarrow \alpha$. This fact can be proved by induction on α. We also have the cut-elimination and completeness (w.r.t. Kripke semantics) theorems for GL. For more information on these theorems, see [30].

Next, we define an intuitionistic Kripke semantics for GL.

Definition 6.3 (Intuitionistic valuation for GL). *An intuitionistic valuation $\models^*$ on a Kripke S4-frame $\langle M, R \rangle$ is a mapping from $\Phi \cup \Phi^\sim$ to 2^M such that for any $p \in \Phi \cup \Phi^\sim$ and any $x, y \in M$, if $x \in \models^* (p)$ and xRy, then $y \in \models^* (p)$. We will write $x \models^* p$ for $x \in \models^* (p)$. This intuitionistic valuation $\models^*$ is extended to a mapping from the set of all formulas to 2^M by:*

1. $x \models^* \alpha \wedge \beta$ iff $x \models^* \alpha$ and $x \models^* \beta$,

2. $x \models^* \alpha \vee \beta$ iff $x \models^* \alpha$ or $x \models^* \beta$,

3. $x \models^* \alpha \to \beta$ iff $\forall y \in M \ [xRy \text{ and } y \models^* \alpha \text{ imply } y \models^* \beta]$,

4. $x \models^* \neg\alpha$ iff $\forall y \in M$ $[xRy$ implies $y \not\models^* \alpha]$,

5. $x \models^* \sim\sim\alpha$ iff $x \models^* \alpha$,

6. $x \models^* \sim(\alpha \wedge \beta)$ iff $x \models^* \sim\alpha$ or $x \models^* \sim\beta$,

7. $x \models^* \sim(\alpha \vee \beta)$ iff $x \models^* \sim\alpha$ and $x \models^* \sim\beta$,

8. $x \models^* \sim(\alpha \to \beta)$ iff $x \models^* \alpha$ and $x \models^* \sim\beta$,

9. $x \models^* \sim\neg\alpha$ iff $x \models^* \alpha$.

Remark 6.4. *The following hereditary condition holds for the intuitionistic valuation $\models^*$. For any formula α and any $x, y \in M$, if $x \models^* \alpha$ and xRy, then $y \models^* \alpha$.*

Definition 6.5 (Kripke GL-model)**.** *A Kripke GL-model is a structure $\langle M, R, \models^* \rangle$ such that*

1. *$\langle M, R \rangle$ is a Kripke S4-frame,*

2. *$\models^*$ is an intuitionistic valuation on $\langle M, R \rangle$.*

A formula α is true *in a Kripke GL-model $\langle M, R, \models^* \rangle$ if $x \models^* \alpha$ for any $x \in M$, and is GL-valid in a Kripke S4-frame $\langle M, R \rangle$ if it is true for every intuitionistic valuation $\models^*$ on the Kripke S4-frame.*

We have the following completeness theorem [21, 23].

Theorem 6.6 (Completeness for GL)**.** *For any formula α, GL $\vdash\ \Rightarrow \alpha$ iff α is GL-valid.*

Proof. Similar to the proof of Theorem 5.19. Namely, this theorem can be proved using the theorems for syntactically and semantically embedding GL into a Gentzen-style sequent calculus LJ for intuitionistic logic. For the detailed proof, see also [23]. □

Next, we introduce a modified Gödel–McKinsey–Tarski translation from GL (Gurevich logic) into $GS4_3$ (S4) and show the Gödel–McKinsey–Tarski theorem using a similar way as presented in [24].

Definition 6.7. *Let $\mathcal{L}_{GL}$ and $\mathcal{L}_{GS4_3}$ be the languages of GL and $GS4_3$, respectively. A mapping g from $\mathcal{L}_{GL}$ into $\mathcal{L}_{GS4_3}$ is defined inductively by:*

1. *for any propositional variable p, $g(p) := \Box p$,*

2. $g(\alpha \sharp \beta) := g(\alpha) \sharp g(\beta)$ where $\sharp \in \{\wedge, \vee\}$,

3. $g(\alpha \to \beta) := \Box(g(\alpha) \to g(\beta))$,

4. $g(\neg \alpha) := \Box \neg g(\alpha)$,

5. for any propositional variable p, $g(\sim p) := \Box \sim p$,

6. $g(\sim \sim \alpha) := g(\alpha)$,

7. $g(\sim(\alpha \wedge \beta)) := g(\sim \alpha) \vee g(\sim \beta)$,

8. $g(\sim(\alpha \vee \beta)) := g(\sim \alpha) \wedge g(\sim \beta)$,

9. $g(\sim(\alpha \to \beta)) := g(\alpha) \wedge g(\sim \beta)$,

10. $g(\sim \neg \alpha) := g(\alpha)$.

Remark 6.8. *The modal operator $\Diamond$ in $\mathcal{L}_{GS4_3}$ is not used in the definition of g in Definition 6.7.*

Lemma 6.9. *Let g be the mapping defined in Definition 6.7. For any formula α in $\mathcal{L}_{GL}$, $GS4_3 \vdash \Box g(\alpha) \Leftrightarrow g(\alpha)$.*

Proof. The case $GS4_3 \vdash \Box g(\alpha) \Rightarrow g(\alpha)$ is obvious. Thus, we show $GS4_3 \vdash g(\alpha) \Rightarrow \Box g(\alpha)$ by induction on the complexity of α. To show this, we also use the fact that $GS4_3 \vdash \Box g(\alpha) \Rightarrow g(\alpha)$. We need to consider the following cases: $\alpha \equiv p$ for any propositional variable p, $\alpha \equiv \beta \wedge \gamma$, $\alpha \equiv \beta \vee \gamma$, $\alpha \equiv \beta \to \gamma$, $\alpha \equiv \neg \beta$, $\alpha \equiv \sim p$ for any propositional variable p, $\alpha \equiv \sim(\beta \wedge \gamma)$, $\alpha \equiv \sim(\beta \vee \gamma)$, $\alpha \equiv \sim(\beta \to \gamma)$, $\alpha \equiv \sim \sim \beta$, and $\alpha \equiv \sim \neg \beta$. We show some cases.

1. $\alpha \equiv \sim p$ for any propositional variable p: We obtain the required fact:

$$\cfrac{\cfrac{\cfrac{\sim p \Rightarrow \sim p}{\Box \sim p \Rightarrow \sim p} (\Box\text{left})}{\Box \sim p \Rightarrow \Box \sim p} (\Box\text{right})}{\Box \sim p \Rightarrow \Box \Box \sim p} (\Box\text{right})$$

where $\Box \sim p \Rightarrow \Box \Box \sim p$ coincides with $g(\sim p) \Rightarrow \Box g(\sim p)$ by the definition of g.

2. $\alpha \equiv \sim(\beta \wedge \gamma)$: By induction hypotheses, we have: $GS4_3 \vdash g(\sim \beta) \Rightarrow \Box g(\sim \beta)$ and $GS4_3 \vdash g(\sim \gamma) \Rightarrow \Box g(\sim \gamma)$. Suppose that S_1 is:

$$g(\sim \beta) \vee g(\sim \gamma) \Rightarrow \Box g(\sim \beta) \vee \Box g(\sim \gamma).$$

Suppose that S_2 is:

$$\Box(\Box g(\sim\beta)\vee\Box g(\sim\gamma))\Rightarrow \Box(g(\sim\beta)\vee g(\sim\gamma)).$$

We then obtain the required fact:

$$\cfrac{\cfrac{\overset{\vdots\ P}{S_1}\quad \Box g(\sim\beta)\vee\Box g(\sim\gamma)\Rightarrow \Box(\Box g(\sim\beta)\vee\Box g(\sim\gamma))}{\cfrac{g(\sim\beta)\vee g(\sim\gamma)\Rightarrow \Box(\Box g(\sim\beta)\vee\Box g(\sim\gamma))}{g(\sim\beta)\vee g(\sim\gamma)\Rightarrow \Box(g(\sim\beta)\vee g(\sim\gamma))}\ \text{(cut)}}\quad \overset{\vdots\ R}{S_2}}{}\ \text{(cut)}$$

where $g(\sim\beta)\vee g(\sim\gamma)$ coincides with $g(\sim(\beta\wedge\gamma))$ by the definition of g, P is of the form:

$$\cfrac{\cfrac{\cfrac{\overset{\vdots\ Ind.hyp.}{g(\sim\beta)\Rightarrow \Box g(\sim\beta)}}{g(\sim\beta)\Rightarrow \Box g(\sim\beta),\Box g(\sim\gamma)}\ \text{(we-right)}}{g(\sim\beta)\Rightarrow \Box g(\sim\beta)\vee\Box g(\sim\gamma)}\ \text{($\vee$right)}\quad \cfrac{\cfrac{\overset{\vdots\ Ind.hyp.}{g(\sim\gamma)\Rightarrow \Box g(\sim\gamma)}}{g(\sim\gamma)\Rightarrow \Box g(\sim\beta),\Box g(\sim\gamma)}\ \text{(we-right)}}{g(\sim\gamma)\Rightarrow \Box g(\sim\beta)\vee\Box g(\sim\gamma)}\ \text{($\vee$right)}}{g(\sim\beta)\vee g(\sim\gamma))\Rightarrow \Box g(\sim\beta)\vee\Box g(\sim\gamma)}\ \text{($\vee$left)},$$

Q is of the form:

$$\cfrac{\cfrac{\overset{\vdots\ Ind.hyp.}{\Box g(\sim\beta)\Rightarrow \Box g(\sim\beta)}}{\cfrac{\vdots\ \text{($\vee$right), (we-right)}}{\cfrac{\Box g(\sim\beta)\Rightarrow \Box g(\sim\beta)\vee\Box g(\sim\gamma)}{\Box g(\sim\beta)\Rightarrow \Box(\Box g(\sim\beta)\vee\Box g(\sim\gamma))}\ \text{($\Box$right${}^\sharp$)}}}\quad \cfrac{\overset{\vdots\ Ind.hyp.}{\Box g(\sim\gamma)\Rightarrow \Box g(\sim\gamma)}}{\cfrac{\vdots\ \text{($\vee$right), (we-right)}}{\cfrac{\Box g(\sim\gamma)\Rightarrow \Box g(\sim\beta)\vee\Box g(\sim\gamma)}{\Box g(\sim\gamma)\Rightarrow \Box(\Box g(\sim\beta)\vee\Box g(\sim\gamma))}\ \text{($\Box$right${}^\sharp$)}}}}{\Box g(\sim\beta)\vee\Box g(\sim\gamma)\Rightarrow \Box(\Box g(\sim\beta)\vee\Box g(\sim\gamma))}\ \text{($\vee$left)},$$

and R is of the form:

$$\cfrac{\cfrac{\cfrac{\overset{\vdots\ Ind.hyp.}{g(\sim\beta)\Rightarrow g(\sim\beta)}}{\Box g(\sim\beta)\Rightarrow g(\sim\beta)}\ \text{($\Box$left)}}{\cfrac{\vdots\ \text{($\vee$right), (we-right)}}{\Box g(\sim\beta)\Rightarrow g(\sim\beta)\vee g(\sim\gamma)}}\quad \cfrac{\cfrac{\overset{\vdots\ Ind.hyp.}{g(\sim\gamma)\Rightarrow g(\sim\gamma)}}{\Box g(\sim\gamma)\Rightarrow g(\sim\gamma)}\ \text{($\Box$left)}}{\cfrac{\vdots\ \text{($\vee$right), (we-right)}}{\Box g(\sim\gamma)\Rightarrow g(\sim\beta)\vee g(\sim\gamma)}}}{\cfrac{\cfrac{\Box g(\sim\beta)\vee\Box g(\sim\gamma)\Rightarrow g(\sim\beta)\vee g(\sim\gamma)}{\Box(\Box g(\sim\beta)\vee\Box g(\sim\gamma))\Rightarrow g(\sim\beta)\vee g(\sim\gamma)}\ \text{($\Box$left)}}{\Box(\Box g(\sim\beta)\vee\Box g(\sim\gamma))\Rightarrow \Box(g(\sim\beta)\vee g(\sim\gamma))}\ \text{($\Box$right${}^\sharp$)}}\ \text{($\vee$left)}.$$

3. $\alpha \equiv \sim(\beta\rightarrow\gamma)$: By induction hypothesis, we have: $\text{GS4}_3 \vdash g(\beta) \Rightarrow \Box g(\beta)$ and $\text{GS4}_3 \vdash g(\sim\gamma) \Rightarrow \Box g(\sim\gamma)$. Suppose that S is

53

$$\Box g(\beta) \wedge \Box g(\sim\gamma) \Rightarrow \Box(g(\beta) \wedge g(\sim\gamma)).$$

Then, we obtain the required fact:

$$\cfrac{\cfrac{\vdots\ P}{\vdots\ S}\quad \cfrac{\cfrac{\cfrac{\cfrac{\cfrac{\cfrac{\cfrac{\vdots\ Prop.\ 2.16}{g(\beta) \Rightarrow g(\beta)}}{g(\beta), g(\sim\gamma) \Rightarrow g(\beta)}\text{(we-left)}\quad \cfrac{\cfrac{\vdots\ Prop..\ 2.16}{g(\sim\gamma) \Rightarrow g(\sim\gamma)}}{g(\beta), g(\sim\gamma) \Rightarrow g(\sim\gamma)}\text{(we-left)}}{g(\beta), g(\sim\gamma) \Rightarrow g(\beta) \wedge g(\sim\gamma)}\text{($\wedge$right)}}{g(\beta), \Box g(\sim\gamma) \Rightarrow g(\beta) \wedge g(\sim\gamma)}\text{($\Box$left)}}{\Box g(\beta), \Box g(\sim\gamma) \Rightarrow g(\beta) \wedge g(\sim\gamma)}\text{($\Box$left)}}{\Box g(\beta), \Box g(\sim\gamma) \Rightarrow \Box(g(\beta) \wedge g(\sim\gamma))}\text{($\Box$right$^\sharp$)}}{\Box g(\beta) \wedge \Box g(\sim\gamma) \Rightarrow \Box(g(\beta) \wedge g(\sim\gamma))}\text{($\wedge$left)}}{g(\gamma) \wedge g(\sim\gamma) \Rightarrow \Box(g(\beta) \wedge g(\sim\gamma))}\text{(cut)},$$

where $g(\beta) \wedge g(\sim\gamma)$ coincides with $g(\sim(\beta\to\gamma))$ by the definition of g, and P is of the form:

$$\cfrac{\cfrac{\cfrac{\vdots\ Ind.hyp.}{g(\beta) \Rightarrow \Box g(\beta)}}{g(\beta), g(\sim\gamma) \Rightarrow \Box g(\beta)}\text{(we-left)}\quad \cfrac{\cfrac{\vdots\ Ind.hyp.}{g(\sim\gamma) \Rightarrow \Box g(\sim\gamma)}}{g(\beta), g(\sim\gamma) \Rightarrow \Box g(\sim\gamma)}\text{(we-left)}}{\cfrac{g(\beta) \wedge g(\sim\gamma) \Rightarrow \Box g(\beta)\quad g(\beta) \wedge g(\sim\gamma) \Rightarrow \Box g(\sim\gamma)}{g(\beta) \wedge g(\sim\gamma) \Rightarrow \Box g(\beta) \wedge \Box g(\sim\gamma)}\text{($\wedge$right)}}\text{($\wedge$left)}.$$

4. $\alpha \equiv \sim\neg\beta$: By induction hypothesis, we have: GS4$_3$ $\vdash g(\beta) \Rightarrow \Box g(\beta)$. Then, we obtain the required fact: GS4$_3$ $\vdash g(\sim\neg\beta) \Rightarrow \Box g(\sim\neg\beta)$ by the definition of g.

$\square$

By using Lemma 6.9, we obtain the following lemma.

Lemma 6.10. *Let g be the mapping defined in Definition 6.7. For any intuitionistic sequent $\Gamma \Rightarrow \gamma$ in $\mathcal{L}_{\text{GL}}$, if GL $\vdash \Gamma \Rightarrow \gamma$, then GS4$_3$ $\vdash g(\Gamma) \Rightarrow g(\gamma)$.*

Proof. By induction on the length of the proofs P of $\Gamma \Rightarrow \gamma$ in GL. We distinguish the cases according to the last inference of P, and show some cases.

1. (init2$^\sharp$): The last inference of P is of the form: $\sim p \Rightarrow \sim p$ for any propositional variable p. We obtain the required fact:

$$\cfrac{\cfrac{\sim p \Rightarrow \sim p}{\Box\sim p \Rightarrow \sim p}\text{($\Box$left)}}{\Box\sim p \Rightarrow \Box\sim p}\text{($\Box$right)}$$

where $\Box\sim p$ coincides with $g(\sim p)$ by the definition of g.

2. (init3$^\sharp$): The last inference of P is of the form: $\sim p, p \Rightarrow$ for any propositional variable p. We obtain the required fact:

$$\frac{\dfrac{\sim p, p \Rightarrow}{\sim p, \Box p \Rightarrow} (\Box\text{left})}{\Box\sim p, \Box p \Rightarrow} (\Box\text{left})$$

where $\Box\sim p$ and $\Box p$ coincide with $g(\sim p)$ and $g(p)$ by the definition of g.

3. ($\rightarrow$right): The last inference of P is of the form:

$$\frac{\alpha, \Gamma \Rightarrow \beta}{\Gamma \Rightarrow \alpha \rightarrow \beta} (\rightarrow\text{right}).$$

By induction hypothesis, we have: $\text{GS4}_3 \vdash g(\alpha), g(\Gamma) \Rightarrow g(\beta)$. We then obtain:

$$\frac{\vdots \ Ind.hyp.}{\dfrac{\dfrac{\dfrac{g(\alpha), g(\Gamma) \Rightarrow g(\beta)}{g(\Gamma) \Rightarrow g(\alpha) \rightarrow g(\beta)} (\rightarrow\text{right})}{\Box g(\Gamma) \Rightarrow g(\alpha) \rightarrow g(\beta)} (\Box\text{left})}{\Box g(\Gamma) \Rightarrow \Box(g(\alpha) \rightarrow g(\beta))} (\Box\text{right}^\sharp)}$$

where $\Box(g(\alpha) \rightarrow g(\beta))$ coincides with $g(\alpha) \rightarrow g(\beta)$ by the definition of g. By Lemma 6.9, we have: $\text{GS4}_3 \vdash g(\gamma) \Leftrightarrow \Box g(\gamma)$ for any $\gamma \in \Gamma$. We then obtain the required fact: $\text{GS4}_3 \vdash g(\Gamma) \Rightarrow g(\alpha \rightarrow \beta)$ by applying (cut) to $g(\gamma) \Rightarrow \Box g(\gamma)$ ($\gamma \in \Gamma$) and $\Box g(\Gamma) \Rightarrow g(\alpha \rightarrow \beta)$, repeatedly.

4. ($\sim\rightarrow$right): The last inference of P is of the form:

$$\frac{\Gamma \Rightarrow \alpha \quad \Gamma \Rightarrow \sim\beta}{\Gamma \Rightarrow \sim(\alpha \rightarrow \beta)} (\sim\rightarrow\text{right}).$$

By induction hypotheses, we have:

$$\text{GS4}_3 \vdash g(\Gamma) \Rightarrow g(\alpha) \text{ and } \text{GS4}_3 \vdash g(\Gamma) \Rightarrow g(\sim\beta).$$

We then obtain the required fact:

$$\frac{\vdots \ Ind.hyp. \qquad \vdots \ Ind.hyp.}{\dfrac{g(\Gamma) \Rightarrow g(\alpha) \qquad g(\Gamma) \Rightarrow g(\sim\beta)}{g(\Gamma) \Rightarrow g(\alpha) \wedge g(\sim\beta)} (\wedge\text{right})}$$

where $g(\alpha) \wedge g(\sim\beta)$ coincides with $g(\sim(\alpha \rightarrow \beta))$ by the definition of g.

5. ($\sim\to$left1): The last inference of P is of the form:

$$\dfrac{\alpha,\Gamma \Rightarrow \gamma}{\sim(\alpha{\to}\beta),\Gamma \Rightarrow \gamma}\ (\sim\to\text{left1}).$$

By induction hypothesis, we have: $\text{GS4}_3 \vdash g(\alpha), g(\Gamma) \Rightarrow g(\gamma)$. We then obtain the required fact:

$$\dfrac{\dfrac{\vdots\ Ind.hyp.}{\dfrac{g(\alpha), g(\Gamma) \Rightarrow g(\gamma)}{\dfrac{g(\alpha), g(\sim\beta), g(\Gamma) \Rightarrow g(\gamma)}{g(\alpha)\wedge g(\sim\beta), g(\Gamma) \Rightarrow g(\gamma)}\ (\wedge\text{left})}\ (\text{we-left})}}{}$$

where $g(\alpha)\wedge g(\sim\beta)$ coincides with $g(\sim(\alpha{\to}\beta))$ by the definition of g.

6. ($\sim\neg$left): The last inference of P is of the form:

$$\dfrac{\alpha, \Gamma \Rightarrow \gamma}{\sim\neg\alpha, \Gamma \Rightarrow \gamma}\ (\sim\neg\text{left}).$$

By induction hypothesis, we have: $\text{GS4}_3 \vdash g(\alpha), g(\Gamma) \Rightarrow g(\gamma)$. Then, we obtain the required fact: $\text{GS4}_3 \vdash g(\sim\neg\alpha), g(\Gamma) \Rightarrow g(\gamma)$ by the definition of g.

□

Lemma 6.11. *Let g be the mapping defined in Definition 6.7. For any formula α in $\mathcal{L}_{\text{GL}}$, if $\text{GS4}_3 \vdash\ \Rightarrow g(\alpha)$, then $\text{GL} \vdash\ \Rightarrow \alpha$.*

Proof. We show that $\text{GL} \nvdash\ \Rightarrow \alpha$ implies $\text{GS4}_3 \nvdash\ \Rightarrow g(\alpha)$. Suppose $\text{GL} \nvdash\ \Rightarrow \alpha$. By Theorem 6.6 (the completeness theorem for GL), there exists a Kripke GL-model $\langle M, R, \models^*\rangle$ such that α is not true in $\langle M, R, \models^*\rangle$. Suppose that $x_0 \models^* \alpha$ does not hold for $x_0 \in M$.

We define an extended Kripke S4-model $\langle M, R, \models^\bullet\rangle$ for GS4_3 by:

1. for any propositional variable p, $x \models^\bullet p$ iff $x \models^* p$,

2. for any propositional variable p, $x \models^\bullet \sim p$ iff $x \models^* \sim p$.

We will show the following claim: For any $\alpha \in \mathcal{L}_{\text{GL}}$ and any $x \in M$,

$$x \models^\bullet g(\alpha)\ \text{iff}\ x \models^* \alpha.$$

By this claim and the assumption ($x_0 \models^* \alpha$ does not hold), we obtain the fact that $x_0 \models^\bullet g(\alpha)$ does not hold. Therefore, GS4$_3$ $\nvdash \Rightarrow g(\alpha)$ by the soundness part of Theorem 5.19 (the completeness theorem for GS4$_3$). Note that it is sufficient to apply the soundness part of the completeness theorem.

We now prove the claim by induction on the complexity of α. We show some cases.

1. $\alpha \equiv \sim p$ for any propositional variable p: We obtain:

    $x \models^* \sim p$
    iff $\forall y \in M \ [xRy \text{ implies } y \models^* \sim p]$ (by the hereditary condition)
    iff $\forall y \in M \ [xRy \text{ implies } y \models^\bullet \sim p]$
    iff $x \models^\bullet \Box \sim p$
    iff $x \models^\bullet g(\sim p)$ (by the definition of g).

2. $\alpha \equiv \beta \to \gamma$: We obtain:

    $x \models^* \beta \to \gamma$
    iff $\forall y \in M \ [xRy \text{ and } y \models^* \beta \text{ imply } y \models^* \gamma]$
    iff $\forall y \in M \ [xRy \text{ and } y \models^\bullet g(\beta) \text{ imply } y \models^\bullet g(\gamma)]$ (by induction hypothesis)
    iff $x \models^\bullet \Box(g(\beta) \to g(\gamma))$
    iff $x \models^\bullet g(\beta \to \gamma)$ (by the definition of g).

3. $\alpha \equiv \sim(\beta \to \gamma)$: We obtain:

    $x \models^* \sim(\beta \to \gamma)$
    iff $x \models^* \beta$ and $x \models^* \sim\gamma$
    iff $x \models^\bullet g(\beta)$ and $x \models^\bullet g(\sim\gamma)$ (by induction hypothesis)
    iff $x \models^\bullet g(\beta) \wedge g(\sim\gamma)$
    iff $x \models^\bullet g(\sim(\beta \to \gamma))$ (by the definition of g).

4. $\alpha \equiv \sim\neg\beta$: We have: $x \models^* \sim\neg\beta$ iff $x \models^* \beta$ iff $x \models^\bullet g(\beta)$ (by induction hypothesis) iff $x \models^\bullet g(\sim\neg\beta)$ (by the definition of g).

□

We obtain the following Gödel–McKinsey–Tarski theorem for embedding GL into GS4$_3$.

Theorem 6.12 (Gödel–McKinsey–Tarski theorem). *Let g be the mapping defined in Definition 6.7. For any formula α in $\mathcal{L}_{\text{GL}}$, GL $\vdash \Rightarrow \alpha$ iff GS4$_3$ $\vdash \Rightarrow g(\alpha)$.*

Proof. By Lemmas 6.10 and 6.11. □

7 Concluding remarks

In this study, we introduced three new Gentzen-style sequent calculi $GS4_1$, $GS4_2$, and $GS4_3$ for normal modal logic S4. The calculi $GS4_1$ and $GS4_2$ were obtained from a Gentzen-style sequent calculus GMA4 for a modal extension MA4 of Avron's self-extensional paradefinite four-valued logic SE4 [7] by adding the rules (explosion) and (ex-middle) or the initial sequents (init3) and (init4), respectively. The calculus $GS4_3$ was obtained by combining $GS4_1$ and $GS4_2$ with the auxiliary negation connective $\sim$. We also introduced the alternative and extended Kripke semantics for $GS4_1$, $GS4_2$, and $GS4_3$. We then proved the cut- and contraposition-elimination and completeness (w.r.t. the alternative and extended Kripke semantics) theorems for $GS4_1$, $GS4_2$, and $GS4_3$. Furthermore, we proved the Gödel–McKinsey–Tarski theorem for embedding a Gentzen-style sequent calculus GL for Gurevich logic [21] into $GS4_3$. Thus, in this study, we clarified the relationship among S4, SE4, and Gurevich logic based on the proposed compatible calculi and the Gödel–McKinsey–Tarski theorem.

In what follows, we address a survey of some closely related works.

First, we explain the study of Avron's logic SE4 and its neighbors because the proposed calculi of the current study are constructed based on SE4 (i.e., the proposed calculi are compatible with SE4). The logic SE4 is known to be a *self-extensional logic* [7], which has *self-extensional property* (also referred to as *replacement property* or *substitution property*). Paradefinite or paraconsistent logics with this property were generally referred to as *self-extensional paradefinite or paraconsistent logics*. In particular, SE4 was shown in [7] to be the unique self-extensional extension of *Belnap–Dunn logic* BD (also referred to as *Dunn–Belnap logic* or *first-degree entailment logic*) [10, 11, 16, 17] by adding implication.

The discovery of self-extensional paradefinite or paraconsistent logics with implication is known to be a significant issue [7] because one of the main drawbacks of many of the well-known paraconsistent logics including *Nelson's paraconsistent four-valued logic* N4 [1, 39, 47, 35, 36] is that they lack this property. Concerned with this issue, it was proved in [8] that no paraconsistent three-valued logic with implication can be self-extensional. This result is a generalization of the results in [12], wherein it was shown that the three-valued logics LP [5, 42] and J3 [18] have no self-extensional property. By contrast, it was proved in [6] that there exists the unique self-extensional paraconsistent three-valued logic within the language $\{\wedge, \vee, \neg\}$. For more information on paraconsistent three-valued logics, see [2]. Concerned with this issue, we proved in the current study that GMA4 (MA4), which is a modal extension of SE4, has the self-extensional property.

It was observed in [25] that a Gentzen-style sequent calculus A4 for SE4 is the classical-negation-free fragment of De and Omori's axiomatic extension BD+ [15] of BD by adding classical negation and classical implication. This means that BD+ is a conservative extension of A4 (SE4) by adding classical negation. The essential equivalence among the following philosophically plausible logics was observed in [15]: *De and Omori's extended Belnap–Dunn logic* BD+, *Béziau's four-valued modal logic* PM4N [13], and *Zaitsev's paraconsistent logic* FDEP [48]. Some Gentzen-style sequent calculi for BD+ and its extensions have also been introduced. A Gentzen-style sequent calculus FBD+ for a first-order extension of BD+ was studied in [33], wherein it was shown that the cut-elimination and completeness theorems for FBD+ hold. A Gentzen-style sequent calculus gBD+ for BD+ was introduced and investigated in [22], wherein the cut-elimination and completeness theorems for gBD+ were proved. It was observed in [25] that the classical-negation-less part of gBD+ is equivalent to A4.

The result of the current study on the Gödel–McKinsey–Tarski theorem for embedding GL into $GS4_3$ is regarded as an analogue of the corresponding result of [24]. In [24], Gentzen-style sequent calculi BDm and BDi were introduced for a modal extension and an intuitionistic modification, respectively, of BD+. The Glivenko theorem for embedding BD+ into BDi and the Gödel–McKinsey–Tarski theorem for embedding BDi into BDm were proved. The results of the current study on GMA4, $GS4_1$ and $GS4_2$ are regarded as modal analogues of the results of [32]. In [32], a first-order extension GA4 of A4 was introduced and two Gentzen-style sequent calculi GCL_1 and GCL_2 for first-order classical logic were constructed by extending GA4. The calculus GCL_1 was obtained from GA4 by adding (explosion) and (ex-middle) and the calculus GCL_2 was obtained from GA4 by adding (init3) and (init4). The results of [31] are regarded as intuitionistic analogues of the results of the current study on GMA4 and $GS4_1$. In [31], a unified and modular falsification-aware single-succedent Gentzen-style framework was introduced for classical, paradefinite, paraconsistent, and paracomplete logics. This framework was composed of the intuitionistic sequent versions (int-explosion) and (int-ex-middle) of (explosion) and (ex-middle), respectively.

References

[1] A. Almukdad and D. Nelson, Constructible falsity and inexact predicates, Journal of Symbolic Logic 49 (1), pp. 231-233, 1984.

[2] O. Arieli and A. Avron, Three-valued paraconsistent propositional logics, In: J.-Y. Béziau, M. Chakraborty, and S. Dutta (eds.), New Directions in Paraconsistent Logic, pp. 91-129, Springer, Berlin, 2015.

[3] O. Arieli and A. Avron, Minimal paradefinite logics for reasoning with incompleteness and inconsistency, Proceedings of the 1st International Conference on Formal Structures for Computation and Deduction (FSCD), Leibniz International Proceedings in Informatics (LIPIcs) 52, pp. 7:1-7:15, 2016.

[4] O. Arieli and A. Avron, Four-valued paradefinite logics, Studia Logica 105 (6), pp. 1087-1122, 2017.

[5] F.G. Asenjo, A calculus of antinomies, Notre Dame Journal of Formal Logic 7, pp. 103-106, 1966.

[6] A. Avron, Self-extensional three-valued paraconsistent logics, Logica Universalis 11 (3), pp. 297-315, 2017.

[7] A. Avron, The normal and self-extensional extension of Dunn–Belnap logic, Logica Universalis 14 (3), pp. 281-296, 2020.

[8] A. Avron and J.-Y. Béziau, Self-extensional three-valued paraconsistent logics have no implication, Logic Journal of the IGPL 25 (2), pp. 183-194, 2017.

[9] A. Avron, O. Arieli, and A. Zamansky, Theory of effective propositional paraconsistent logics, Studies in Logic - Mathematical Logic and Foundations, Volume 75, 550 pages, College Publications, 2018.

[10] N.D. Belnap, A useful four-valued logic, In: Modern Uses of Multiple-Valued Logic, G. Epstein and J. M. Dunn (eds.), Dordrecht: Reidel, pp. 5-37, 1977.

[11] N.D. Belnap, How a computer should think, In: Contemporary Aspects of Philosophy, G. Ryle (ed.), Oriel Press, Stocksfield, pp. 30-56, 1977.

[12] J.-Y. Béziau, Idempotent full paraconsistent negations are not algebraizable, Notre Dame Journal of Formal Logic 39, pp. 135-139, 1994.

[13] J.-Y. Béziau, A new four-valued approach to modal logic, Logique et Analyse 54 (213), pp. 109-121, 2011.

[14] J.-Y. Béziau, Bivalent semantics for De Morgan logic (The uselessness of four-valuedness), In W.A. Carnielli, M.E. Coniglio, and I.M. D'Ottaviano (eds.), The many sides of logic, pp. 391-402, College Publications, 2009.

[15] M. De and H. Omori, Classical negation and expansions of Belnap-Dunn logic, Studia Logica 103 (4), pp. 825-851, 2015.

[16] J.M. Dunn, Intuitive semantics for first-degree entailment and 'coupled trees', Philosophical Studies 29 (3), pp. 149-168, 1976.

[17] J.M. Dunn, The Algebra of Intensional Logics, A republication of a PhD thesis, which was originally defended at the University of Pittsburgh in 1966, College Publications, London, UK, 2019.

[18] I. D'Ottaviano, The completeness and compactness of a three-valued first-order logic, Revista Colombiana de Matemáticas, XIX (1-2), pp. 31-42, 1985.

[19] G. Gentzen, Collected papers of Gerhard Gentzen, M.E. Szabo (ed.), Studies in logic and the foundations of mathematics, North-Holland (English translation), 1969.

[20] O. Grigoriev and Y. Petrukhin, On a multilattice analogue of a hypersequent S5 calculus, Logic and Logical Philosophy 28, pp. 683-730, 2019.

[21] Y. Gurevich, Intuitionistic logic with strong negation, Studia Logica 36, pp. 49-59, 1977.

[22] N. Kamide, Gentzen-type sequent calculi for extended Belnap–Dunn logics with classical negation: A general framework, Logica Universalis 13 (1), pp. 37-63, 2019.

[23] N. Kamide, Cut-elimination, completeness, and Craig interpolation theorems for Gurevich's extended first-order intuitionistic logic with strong negation, Journal of Applied Logics 8 (5), pp. 1101-1122, 2020.

[24] N. Kamide, Modal and intuitionistic variants of extended Belnap–Dunn logic with classical negation, Journal of Logic, Language and Information 30 (3), pp. 491-531, 2021.

[25] N. Kamide, Notes on Avron's self-extensional four-valued paradefinite logic, Proceedings of the 51st IEEE International Symposium on Multiple-Valued Logic (ISMVL 2021), pp. 43-49, IEEE Press, 2021.

[26] N. Kamide, Falsification-aware semantics and sequent calculi for classical logic, Journal of Philosophical Logic 51 (1), pp. 99-126, 2022.

[27] N. Kamide, Herbrand and contraposition-elimination theorems for extended first-order Belnap-Dunn logic, In: Relevance Logics and other Tools for Reasoning: Essays in Honor of J. Michael Dunn (Katalin Bimbo editor), Volume 46 of Tribute Series, pp. 237-260, College Publications, 2022.

[28] N. Kamide, Falsification-aware calculi and semantics for normal modal logics including S4 and S5, Journal of Logic, Language, and Information 32 (3), pp. 395-440, 2023.

[29] N. Kamide, Self-extensional paradefinite four-valued modal logic compatible with standard modal logic, Proceedings of the 53rd IEEE International Symposium on Multiple-Valued Logic (ISMVL 2023), pp. 30-35, 2023.

[30] N. Kamide, Embedding first-order classical logic into Gurevich's extended first-order intuitionistic logic: The role of strong negation, Journal of Applied Logics 10 (6), pp. 1025-1058, 2023.

[31] N. Kamide, Rules of explosion and excluded middle: Constructing a unified single-succedent Gentzen-style framework for classical, paradefinite, paraconsistent, and paracomplete logics, Journal of Logic, Language and Information, 33(2), pp. 143-178, 2024.

[32] N. Kamide, From first-order self-extensional paradefinite four-valued logic to first-order classical logic: Cut-elimination, completeness and self-extensionality, Draft, 32 pages, 2022.

[33] N. Kamide and H. Omori, An extended first-order Belnap–Dunn logic with classical negation, Proceedings of the 6th International Workshop on Logic, Rationality, and Interaction (LORI 2017), Lecture Notes in Computer Science 10455, pp. 79-93, 2017.

[34] N. Kamide and H. Wansing, Symmetric and dual paraconsistent logics, Logic and Logical Philosophy 19 (1-2), pp. 7-30, 2010.

[35] N. Kamide and H. Wansing, Proof theory of Nelson's paraconsistent logic: a uniform perspective, Theoretical Computer Science 415, pp. 1-38, 2012.

[36] N. Kamide and H. Wansing, Proof theory of N4-related paraconsistent logics, Studies in Logic, Volume 54, College Publications, pp. 1-401, 2015.

[37] S. A. Kripke, Semantical analysis of modal logic I Normal modal propositional calculi, Zeitschr. math. Logik und Grundlagen d. Math. Bd. 9, S. pp. 67-96, 1963.

[38] S. Negri and J. von Plato, Structural proof theory, Cambridge University Press, 2001.

[39] D. Nelson, Constructible falsity, Journal of Symbolic Logic 14, pp. 16-26, 1949.

[40] M. Ohnishi and K. Matsumoto, Gentzen method in modal calculi, Osaka Mathematical Journal 9, pp. 113-130, 1957.

[41] M. Ohnishi and K. Matsumoto, Gentzen method in modal calculi II, Osaka Mathematical Journal 11, pp. 115-120, 1959.

[42] G. Priest, The logic of paradox, Journal of Philosophical Logic 8 (1), pp. 219-241, 1979.

[43] G. Priest, Paraconsistent logic, Handbook of Philosophical Logic (Second Edition), Vol. 6, D. Gabbay and F. Guenthner (eds.), Kluwer Academic Publishers, Dordrecht, pp. 287-393, 2002.

[44] W. Rautenberg, Klassische und nicht-klassische Aussagenlogik, Vieweg, Braunschweig, 1979.

[45] J. von Plato, Proof theory of full classical propositional logic, Draft, 16 pages, 1998.

[46] N.N. Vorob'ev, A constructive propositional calculus with strong negation (in Russian), Doklady Akademii Nauk SSSR 85, pp. 465-468, 1952.

[47] H. Wansing, The logic of information structures, Lecture Notes in Computer Science 681, 163 pages, Springer 1993.

[48] D. Zaitsev, Generalized relevant logic and models of reasoning, Moscow State Lomonosov University doctoral dissertation, 2012.

Intermediate-qudit Assisted Improved Quantum Algorithm for String Matching with an Advanced Decomposition of Fredkin Gate

Amit Saha
Atos, Pune, India
abamitsaha@gmail.com

Om Khanna
Atos, Pune, India, and Jai Hind College (Autonomous), University of Mumbai, Mumbai, India
aumkhanna@gmail.com

Abstract

The string-matching problem has a broad variety of applications due to its pattern-matching ability. The circuit-level implementation of a quantum string-matching algorithm, which matches a search string (pattern) of length M inside a longer text of length N, has already been demonstrated in the literature to outperform its classical counterparts in terms of time complexity and space complexity. Higher-dimensional quantum computing is becoming more and more common as a result of its powerful storage and processing capabilities. In this article, we have shown an improved quantum circuit implementation for the string-matching problem with the help of higher-dimensional intermediate temporary qudits. It is also shown that with the help of intermediate qudits not only the complexity of depth can be reduced but also query complexity can be reduced for a quantum algorithm, for the first time to the best of our knowledge. Our algorithm has an improved query complexity of $O(\sqrt{N-M+1})$ with overall time complexity $O\left(\sqrt{N-M+1}((\log{(N-M+1)}\log N) + \log(M))\right)$ as compared to the state-of-the-art work which has a query complexity of $O(\sqrt{N})$ with overall time complexity $O\left(\sqrt{N}((\log N)^2 + \log(M))\right)$, while the ancilla count also reduces to $\frac{N}{2}$ from $\frac{N}{2} + M$. The cost of the state-of-the-art quantum circuits for string-matching problem is colossal due to a huge number of Fredkin gates and multi-controlled Toffoli gates. We have exhibited an improved gate cost and depth over the circuit by applying a proposed Fredkin gate decomposition with intermediate qutrits (3-dimensional qudits or ternary systems) and already existing logarithmic-depth decomposition of n-qubit Toffoli or multi-controlled Toffoli gate (MCT) with intermediate ququarts (4-dimensional qudits or quaternary systems). We have also asserted that the quantum circuit cost is relevant instead of using higher dimensional qudits through error analysis.

1 Introduction

Quantum entanglement and superposition are two examples of quantum mechanical phenomena that are used in the idea of quantum computing for an asymptotic advantage [,]. While the fundamental physics of quantum systems is not inherently binary, quantum computation is frequently stated as a two-level binary abstraction of qubits. However, higher dimensional systems can also be used to describe quantum processing. A qubit is expanded to a d-level or d-dimensional structure as a qudit [,]. In this article, an asymptotically improved binary circuit implementation of string-matching problem [ ], has been addressed with temporary intermediate qutrits and ququarts by efficient decomposition of Fredkin gate [ ]. Since these only exist as intermediary states in a qudit system, where the input and output states are qubits, we can readily create a higher dimensional quantum state for temporary use by adding a distinct energy level [ ].

An essential family of algorithms known as "string-matching algorithms" looks for the location of one or more strings (also known as patterns) within a larger string or text. These algorithms are used to discover answers for problems like text mining, pattern recognition, document matching, information security, network intrusion detection, and plagiarism detection. When using exact matching, the pattern is precisely located within the text. The brute force algorithm is the most basic type of algorithm for finding a precise match in the string-matching problem. For example String $\mathcal{T}$ (to be searched) = ABCDEFGH and Pattern, $\mathcal{P}$ (to be matched) = CDEFG and P occurs once in $\mathcal{T}$: ABCDEFGH. With the brute force method, we merely attempt to match the first character of the pattern with the first character of the text. If we are successful, we move on to the second character, and so forth. We move the pattern over one letter and attempt again if we run into a failure point. As a result, this method runs in $O(nm)$ time. However, the Knuth-Pratt-Morris algorithm, which has a worst-case temporal complexity of $\Theta(N+M)$, is the most well-known classical string-matching algorithm [ ]. The most popular approximate string-matching algorithm also has a comparable run-time of $\Theta(N+M)$ [ ].

Quantum computing can be used to speed up string-matching algorithms. A precise string-matching quantum algorithm with $\tilde{O}(\sqrt{N} + \sqrt{M})$ query complexity was developed by Ramesh and Vinay [ ]. In this method, each check is made using a nested Grover search to determine the location where a section of length M from $\mathcal{T}$ matches the pattern $\mathcal{P}$. However, this work does not create the specific oracles needed, and once we take into consideration the gate-level complexity of getting the text and pattern from a database, the total time complexity, expressed in units of gate depth, is bound to rise. For average-case matching, a different strategy for the dihedral hidden subgroup problem [ ] has a time complexity of $\tilde{O}\left((N/M)^{1/2} 2^{O(\sqrt{\log(M)})}\right)$ [ ]. The state-of-the-art work [ ] presents a string-matching algorithm, based on generalized Grover's amplitude amplification [ ], with time complexity of $O\left(\sqrt{N}\left((\log N)^2 + \log(M)\right)\right)$ along with $\frac{N}{2} + M$ ancilla for arbitrary text length N and pattern length $M \leq N$. In this particular paper, we are also using the Grover-based string-matching algorithm to solve the string-matching problem, which achieves time complexity of $O\left(\sqrt{N - M + 1}\left((\log(N - M + 1)\log N) + \log(M)\right)\right)$ with $\frac{N}{2}$ ancilla. We are using a system of intermediate qudits to implement a circuit that provides an asymptotic advantage over the state-of-the-art algorithm.

The main contribution of the article is summarized below:

- We exhibit a first-of-its-kind approach to implement an improved algorithm for the string-matching problem using a novel proposed decomposition of Fredkin gate using intermediate qutrit and multi-controlled Toffoli decomposition with intermediate ququart.

- The proposed approach is sublimer with respect to the time complexity and space complexity with reduced ancilla qubits as compared to the state-of-the-art approach [ ].

- Our approach of solving string-matching problem outperforms the state-of-the-art approach [ ] with respect to circuit cost.

This paper has the following format. The background research required to carry out this suggested work is covered in Section 2. The circuit construction for the string-matching algorithm is proposed in Section 3 by decomposing the Fredkin gate using intermediate qutrits. The efficacy of the suggested approach in comparison to the state-of-the-art is analyzed in section 4. Our findings are summarized in Section 5.

2 Preliminaries

2.1 A State-of-the-art Quantum Algorithm for String-matching

The primary objective of string-matching algorithms is to find the location of a specific text pattern (P) within a larger string (S). The practical importance of these algorithms is in a wide variety of applications, from something as simple as searching for a particular word in a word processor to mapping DNA.

In string-matching, we are given a long string S of length N, and our goal is to search for a pattern P contained in the string of length M, such that $M \leq N$. In Pradeep and Yunseong's state-of-the-art paper [ ], they constructed a quantum string-matching algorithm with a time complexity of $O(\sqrt{N}((\log(N))^2 + \log(M))))$. The steps involved in their algorithm are as follows:

1. It is based on the generalized Grover's amplitude amplification technique. It works by initializing 2 quantum registers to store the bits of the target string of length N and the pattern of length M. This process is done by using the identity and bit flip gates on 2 quantum registers ($|t_0 t_1 t_2 \ldots t_{N-1}\rangle |p_0 p_1 \ldots p_{M-1}\rangle$, where t_i and p_i denote the ith bit of string $\mathcal{T}$ and pattern $\mathcal{P}$, respectively).

2. The first register that contains the string $\mathcal{T}$ is changed into a combination of N states, each of which is a bit-shifted version of the first register's initial state that has been moved by 0, 1, 2,..., $N-1$ bits. As a consequence, and presuming that the bit indices are stored in modulo-N space,

$$\left(\frac{1}{\sqrt{N}} \sum_{k=0}^{N-1} |t_{0+k} t_{1+k} t_{2+k} \ldots t_{N-1+k}\rangle\right) |p_0 p_1 \ldots p_{M-1}\rangle$$

This is done by a cyclic shift operator S and the decomposition of the cyclic shift operator's circuit is shown in Figure 1.

3. Then, the XOR operation is performed on the first M bits of the first register and the entire M bits of the second register to obtain

$$\frac{1}{\sqrt{N}} \sum_k |t_{0+k} t_{1+k} \ldots t_{N-1+k}\rangle$$
$$|(p_0 \oplus t_{0+k})(p_1 \oplus t_{1+k}) \ldots (p_{M-1} \oplus t_{M-1+k})\rangle .$$

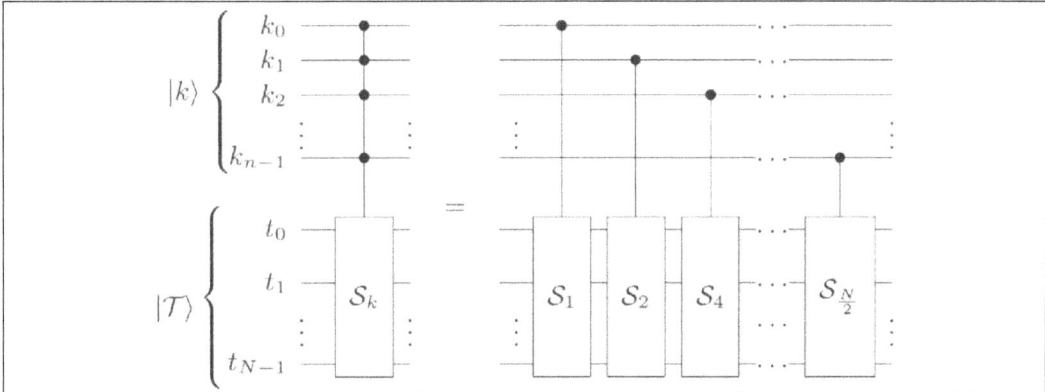

Figure 1: Circuit construction of cyclic shift operator [ ].

4. If the sequence matches the first M bits of the $\mathcal{T}$, the second register contains only zeroes. If the text and the pattern vary by d bit positions, the register holds d ones.

5. When looking for an exact match, the state where the second register contains only zeros or has fewer than D matches using the generalized Grover search or amplitude amplification (in the case of fuzzy search), should be separated.

A comparison of query and time-complexity between our work and other works [, ,] is given in Table 1. The oracles for [,] offer arbitrary access to text and pattern bits. Because the execution time relies on the random-access oracles, which don't have a circuit-level design in the relevant papers, the time complexity for [,] is unclear. In [ ] and our work, this random-access generator is not necessary. Instead, for the purposes of our work, an oracle is a Grover oracle that determines whether a register is in an all-zero state, similar to [ ]. Our detailed construction for such an oracle is discussed in this paper. We also follow the same algorithmic steps as [ ]. Albeit we design our circuit in such a way that we achieve an asymptotic advantage over [ ] with the help of intermediate temporary qudits. Hence we directly compare our time complexity with [ ]. The time complexity consists of three different parts, the first is for the query, the next is for the depth of the cyclic shift operator and the final part is for Grover's search. From Table 1, it can be visualized that our proposed approach has an asymptotic advantage for the first two parts i.e., $\sqrt{N-M+1}$ and $\log{(N-M+1)}\log N$ as compared to $\sqrt{N}$ and $(\log N)^2$. For the last part, the complexity remains the same for the two approaches, but ancilla reduces to 0 from M.

Paper	Query complexity	Time complexity
[ ]	$O\left(\sqrt{N}\log(\sqrt{N/M})\log M + \sqrt{M}(\log M)^2\right)$	-
[ ]	$O\left((\sqrt{N/M})2^{(3/2)}\sqrt{(2\log_2 3)\log_2 M}(\log M)^{3/2}\log N\right)$	-
[ ]	$O(\sqrt{N})$	$O\left(\sqrt{N}\left((\log N)^2 + \log(M)\right)\right)$
This work	$O(\sqrt{N-M+1})$	$O\left(\sqrt{N-M+1}\left(\log(N-M+1)\log N\right) + \log(M)\right)$

Table 1: Comparison of our work with prior algorithms discussed [, ,].

2.2 Toffoli Decomposition via Intermediate Qutrits

Natural access to an infinite range of discrete energy levels is available to quantum processors. Therefore, using three-level qutrits is just an option to add another distinct energy level, but at the expense of allowing more space for error. Qutrits can replace the workspace provided by non-data ancilla qubits in typical circuits, allowing us to function more effectively. Qutrits are a 3-level quantum system where we consider the computational basis states: $|0\rangle$, $|1\rangle$, and $|2\rangle$. They are manipulated similarly to qubits, however, there are additional ternary CNOT gates that may be performed on qutrits during the Toffoli decomposition. The Toffoli gate is the central building block of several quantum algorithms. Since the Toffoli involves 3-body interactions, it cannot be implemented naturally in real quantum devices. Usually, the Toffoli gate can be constructed by decomposing it into single and two-qubit Clifford+T gates. For example, CNOT gates require 6 such gates plus 7 T gates [,] as shown in Fig. 2. Let's look at the decomposition of the Toffoli gate using intermediate qutrits, since in this paper we are using Toffoli decomposition with intermediate qutrit:

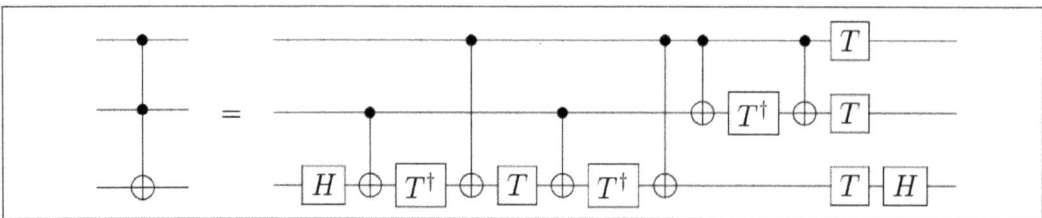

Figure 2: Qubit-only Toffoli decomposition with Clifford+T gate-set [,].

In [ ], the authors demonstrated that we can momentarily inhabit the $|2\rangle$ state during the computation, making it temporarily ternary. This circuit design can be integrated into any current qubit-only circuits because it maintains binary input and output. Fig. 3 depicts a Toffoli realization as seen through qutrits [ ]. More precisely, the target qubit (third qubit) must undergo a NOT operation as long as the two control qubits are both $|1\rangle$. The first and second qubits are then subjected to a $|1\rangle$-controlled X_3^{+1}, where $+1$ stands for an increase of 1 (mod3) to the target qubit. If and only if the first and second qubits were both $|1\rangle$, this raises the second qubit to $|2\rangle$. The target qubit is then subjected to a X gate that is regulated by $|2\rangle$. As anticipated, X is only performed when the first and second qubits were both $|1\rangle$. Lastly, a $|1\rangle$-controlled X_3^{-1} gate cancels the impact of the first gate, returning the controls to their initial positions. The main result of this reduction is that the transient information can be stored in the $|2\rangle$ state from ternary quantum systems instead of ancilla. Therefore, three generalized ternary CNOT gates with a circuit depth of three are adequate to realize the Toffoli gate, in actuality, no T gate is needed.

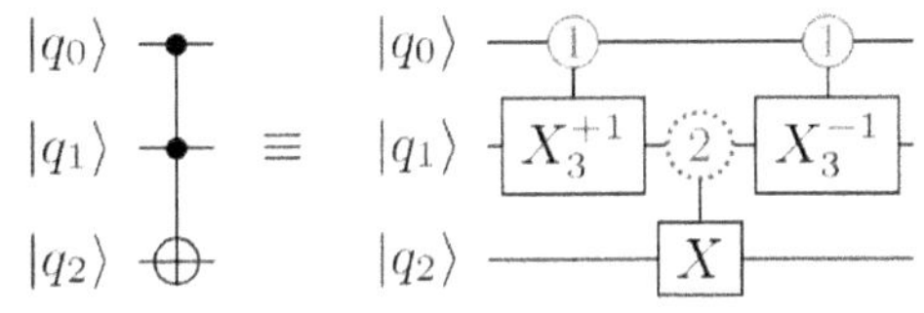

Figure 3: An example of Toffoli decomposition with intermediate qutrit, where input and output are qubits. The red controls activate on $|1\rangle$ and the blue controls activate on $|2\rangle$. The first gate temporarily elevates q_1 to $|2\rangle$ if both q_0 and q_1 were $|1\rangle$. X operation is then only performed if q_1 is $|2\rangle$. The final gate acts as a mirror of the first gate and restores q_0 and q_1 to their original state [ ]

This Toffoli decomposition is further used to decompose the Fredkin gate for string-matching, which is thoroughly discussed in the next section. Before that, we showcased the decomposition of a multi-controlled Toffoli gate using ququarts, which is another important fundamental component for Grover's based string-matching.

2.3 Multi-controlled Toffoli Decomposition via Intermediate Ququarts

In the previous section, we dealt with the construction of a Toffoli gate using a 3-level quantum system i.e., an intermediate qutrit. For the decomposition of n-qubit Toffoli gate, the resources increase rapidly, requiring $O(n^2)$ two-qubit gates in qubit-only systems. However, n-qubit Toffoli gates can be constructed efficiently using fewer resources than previous qubit-only designs with the help of intermediate qutrits [ ]. Similar to this, there is research on the realization of n-qubit Toffoli with intermediary qudits; see [, , ,]. Since this decomposition [ ] is more error-resistant and is the only one that can be scaled up to any finite-dimensional quantum system as opposed to [, ,], we are using it in this paper. It should be mentioned that the circuit cost of decomposition with intermediate qutrits [ ] and the decomposition with intermediate ququarts [ ] is comparable. Even so, not all quantum hardware supports the used gate-set from [ ], and there is no error analysis for this decomposition because the error rates for the used gate-set and ternary systems are not documented in the literature. The gate-set utilised in [ ] is also not scalable to any finite-dimensional system because it is not generalized to any such system.

As an illustration, a multi-controlled Toffoli gate with 7 control qubits and 1 target qubit is taken into consideration, as shown in Fig. 4(a). With the support of the Gokhale et al. [ ] design, Fig. 4(b) shows the realization of the generalized 8-qubit Toffoli gate as shown in Fig. 4(a). In the same way that this method does, their circuit briefly saves information in the qutrit $|2\rangle$ state of the controls. However, they decompose their ternary Toffoli into 13 one-qutrit and two-qutrit gates, [ ] [ ], rather than saving temporary states in the quaternary $|3\rangle$ state. According to this method, three ternary and/or quaternary CNOT gates can be further reduced to two ternary and/or quaternary CNOT gates by using the identity rule, as shown in Fig. 4(c) on Fig. 4(d). The authors further decompose the ternary Toffoli into three ternary and/or quaternary CNOT gates using the $|3\rangle$. As a result, for a single Toffoli decomposition, this optimization can reduce the gate count from 13 to 2, and this method can also be applied to any dimensional quantum system.

This circuit design, as displayed in 4(c) or 4(d), can be understood as a binary tree of gates. More specifically, the circuit retains a tree structure with qubit inputs and outputs, and it has the characteristic that the intermediate qubit of each sub-tree and root can only be raised to $|2\rangle$ if all seven of its control leaves were $|1\rangle$. As a result, the circuit depth, where n is the total number of controls, is exponential in n. Additionally, the overall number of gates is optimized because each quaternary qudit is acted on by a small constant number of two gates. The n-qubit Toffoli decomposition is novel because it uses a maximum of $2n-3$ generalized CNOT gates ($n+1$ ternary CNOT gates and $n-4$ quaternary CNOT gates), which is less than the state-of-the-art. It is also novel because of its logarithmic depth optimization. This decomposition of the multi-controlled Toffoli gate has further played a vital role in reducing the query complexity and ancilla qubits of our proposed string-matching algorithm, which is discussed in the next section.

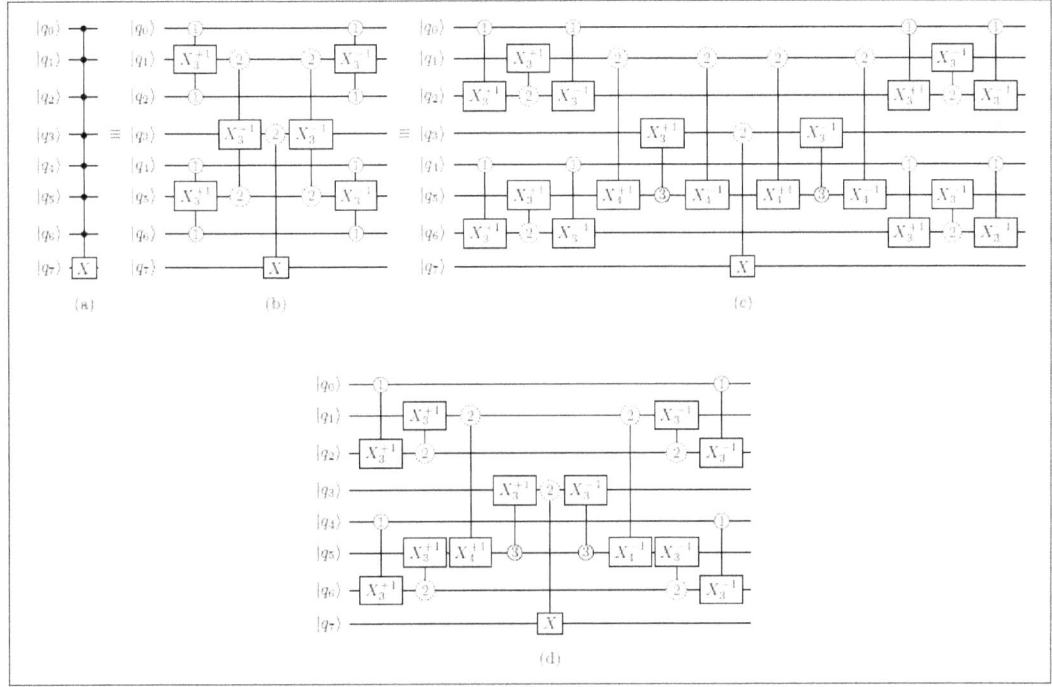

Figure 4: (a) An 8-qubit Toffoli gate, (b) its decomposition in [ ], (c) its decomposition using a few ternary and/or quaternary CNOT gates in [ ], and (d) its optimized decomposition in [ ].

3 Quantum Algorithm for String-matching with Intermediate Qudits

3.1 Our Proposed Fredkin Gate with Intermediate Qutrits

In this section, we show an explicit circuit decomposition of the Fredkin gate using intermediate qutrits. Before that, the state-of-the-art decomposition of a Fredkin gate with 7 CNOT and 7 T gates is discussed. Suppose we start with the circuit of Fredkin gate:

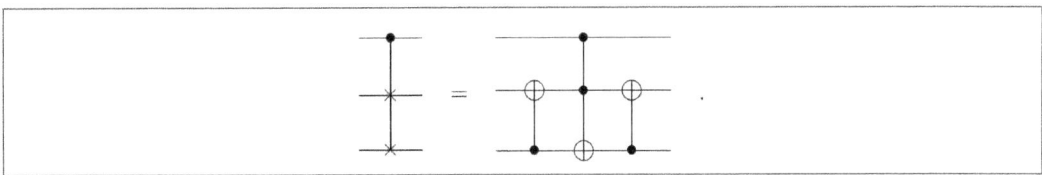

Using the circuit identity, imported from Fig. 2 of [ ], we find that the first CNOT gate in the Fredkin-gate circuit and the first two gates of the Toffoli-gate circuit form a subcircuit. Thus, we obtain the state-of-the-art decomposition of Fredkin gate with Clifford +T gate set:

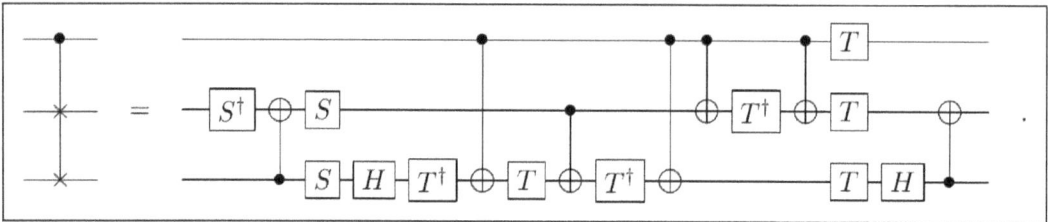

The proposed Fredkin gate with intermediate qutrit is shown in Figure 5, where the Toffoli gate is decomposed as per Fig. 3. Only 2 CNOT gates and 3 ternary CNOT gates are required to construct this Fredkin gate, in fact, no T gate is required. We use this Fredkin gate further in our proposed string matching algorithm.

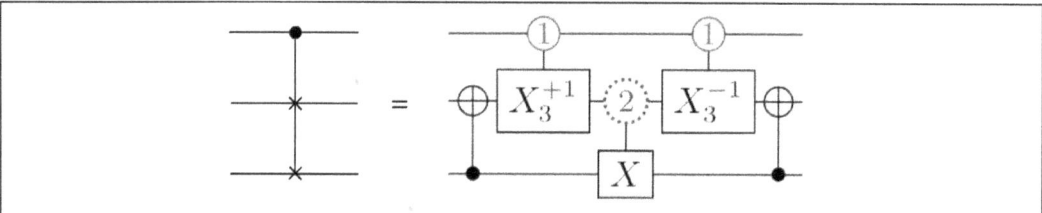

Figure 5: Advanced Fredkin gate with intermediate qutrit.

3.2 Our Proposed Methodology for String-matching using Grover's Algorithm with Proposed Fredkin Gate

We outline the proposed algorithm's thorough implementation in this section. We specifically describe the registers and transformations used to carry out the method. The cyclic shift operator with the proposed Fredkin gate is one of the key changes that will be used in our method compared to [ ]. Another key aspect of our algorithm is that due to the use of multi-controlled Toffoli decomposition with intermediate qudits, the query complexity has been reduced. We present the details of complete circuit construction for string-matching using Grover's algorithm, which is portrayed in Fig. 6 for better visualization.

Step 1: We also use quantum registers of N and M qubits, respectively, to encapsulate a binary string $\mathcal{T}$ of length N and a binary pattern $\mathcal{P}$ of length M as [ ]. To

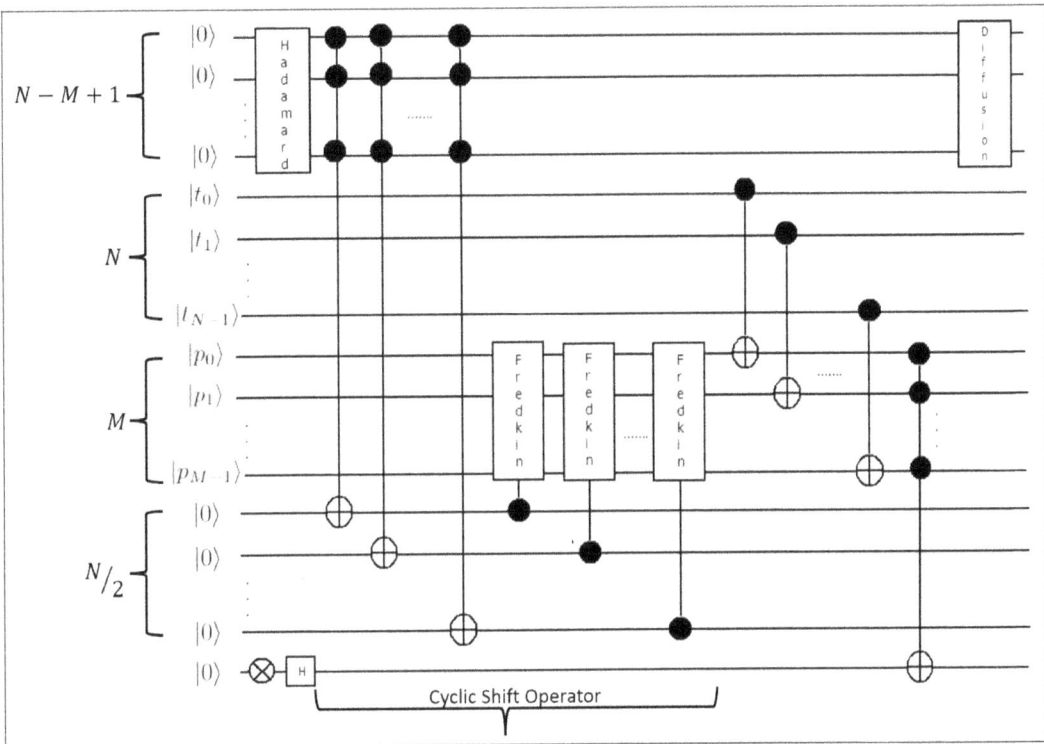

Figure 6: Complete circuit for proposed string-matching.

accomplish this, identity and bit-flip gates can be used on a quantum register with an initialization of $|0\rangle^{\otimes(N+M)}$. The encoded quantum state is as follows

$$|\mathcal{T}\rangle = |t_0 t_1 \ldots t_{N-1}\rangle = \bigotimes_{i=0}^{N-1} |t_i\rangle$$

$$|\mathcal{P}\rangle = |p_0 p_1 \ldots p_{M-1}\rangle = \bigotimes_{j=0}^{M-1} |p_j\rangle.$$

Step 2: We now construct a composite initial state and an index register of $N - M + 1$ qubits in the zero states,

$$|\psi\rangle = |0\rangle^{\otimes N-M+1} \left[\bigotimes_{i=0}^{N-1} |t_i\rangle \right] \left[\bigotimes_{j=0}^{M-1} |p_j\rangle \right]$$

where, for ease of use, we assumed $N - M + 1 = 2^n$. The index register is then subjected to a n qubit Hadamard transform $H^{\otimes n}$ (Fourier transform in case of $N - M + 1 \neq 2^n$ for $n \in \mathbb{N}$) to produce a uniform superposition of $|0\rangle, |1\rangle, \ldots |N - M\rangle$,

$$(H^{\otimes n}|0\rangle^{\otimes n}) \left[\bigotimes_{i=0}^{N-1} |t_i\rangle \right] \left[\bigotimes_{j=0}^{M-1} |p_j\rangle \right]$$

$$= \left(\frac{1}{\sqrt{N-M+1}} \sum_{k=0}^{N-M} |k\rangle \right) \left[\bigotimes_{i=0}^{N-1} |t_i\rangle \right] \left[\bigotimes_{j=0}^{M-1} |p_j\rangle \right].$$

Step 3: The next step is to use the cyclic shift operator $\mathcal{S}$, which left-circularly shifts the target state's qubits by k places. k's values are stored in the control state. The outcome of applying $\mathcal{S}$ to the first two registers is

$$\left[\mathcal{S} \left(\frac{1}{\sqrt{N-M+1}} \sum_{k=0}^{N-M} |k\rangle \right) \left(\bigotimes_{i=0}^{N-1} |t_i\rangle \right) \right] \left(\bigotimes_{j=0}^{M-1} |p_j\rangle \right)$$

$$= \frac{1}{\sqrt{N-M+1}} \sum_{k=0}^{N-M} |k\rangle \left(\bigotimes_{i=0}^{N-1} |t_{i+k}\rangle \right) \left(\bigotimes_{j=0}^{M-1} |p_j\rangle \right)$$

Here, we provide a short explanation of the circuit design for the $\mathcal{S}$ cyclic-shift operator. We consider k in its binary encoded form $|k_0\rangle |k_1\rangle \ldots |k_{N-M}\rangle$, such that $2^0 k_0 + 2^1 k_1 + \ldots + 2^{N-M} k_{N-M} = k$, to implement the k-controlled circular shift operator S_k. Then, a combination of controlled-shift operators that shifts the target qubits by k bits while depending on the k-controlled qubits can be used to execute the circular bitwise shift by k in the second register. In other terms, a product of controlled shift operations can produce a shift of k bits. We need the controlled-SWAP (Fredkin) gates to put the circular shift operator into practice. As an instance, a permutation of the form $P_r = \{N - r, N - r + 1, N - r + 2, \ldots, N - r - 1\}$ is applied in modulo N space by applying a cyclic shift operator S_r by r bits, where the $N - r$th bit is inserted in the zeroth position, the $N - r + 1$th bit is inserted in the first position, and so on. Any one of these permutations can be realized into a series of transpositions. As a consequence, a cyclic shift operation can be realized into a SWAP operation's byproduct.

The number of SWAP-operation levels required to effectively implement the permutation is now determined. With a register having N qubits, we can perform $\frac{N}{2}$ SWAP processes in parallel. We can transfer $\frac{N}{2}$ qubits to the appropriate locations in a single time step by using the $\frac{N}{2}$-parallel SWAP operator. Now we just need to arrange the remaining $N/2$ bits. The number of qubits that must be swapped drops by half at each succeeding time step. Therefore, using concurrent SWAP operations, we can arbitrarily permute N qubits in $O(\log(N))$ time steps. This unitary process is illustrated diagrammatically with an example in Fig. 7. By using concurrent controlled-SWAP operators, shift operators can be implemented in $O(\log(N))$ time steps.

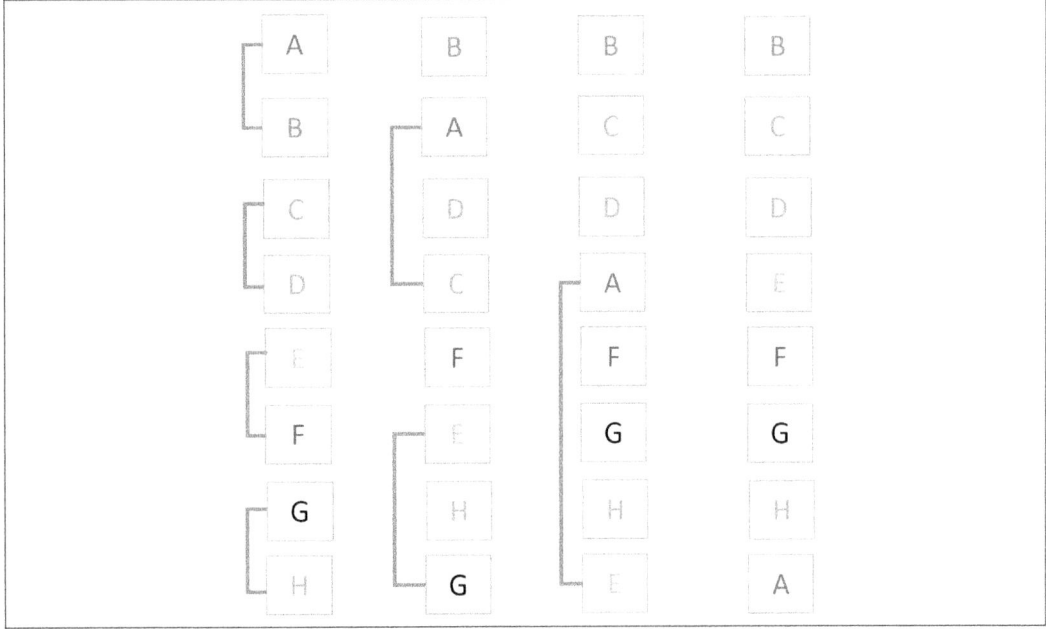

Figure 7: The cyclic-shift operator is shown in this diagram. In this case, we left-circularly shifted an 8-qubit register by one location over the course of three-time steps. Generally speaking, this type of procedure can be carried out in depth $\log(N)$ using parallel SWAP operations, where N is the size of the qubit register states.

Next, we go over how to use the same qubits in the index register to handle as many $\frac{N}{2}$ parallel swap processes. We succeed in doing this, at the expense of $\frac{N}{2}$ clean ancilla qubits. We start by considering an MCT operation, acting on the control qubits in a state $|k\rangle$ and $\frac{N}{2}$ clean ancilla qubits initialized to $|0\rangle$ as targets. This results in $\frac{N}{2}$ copies of $|1\rangle$, which can then be used to implement up to $\frac{N}{2}$ Fredkin gates in a single time step. Once all necessary Fredkin gates have been implemented, we undo the MCT operation and return all ancilla qubits to $|0\rangle$ for further operations. The time cost of the MCT operations with intermediate qudit is $O(\log(N - M + 1))$ as index register of $\lceil \log(N - M + 1) \rceil$ qubits are enough for our string matching since $N - M + 1$ time cyclic shift operator is needed to be performed. For [ ], they need an index register of N qubits as the logarithmic decomposition of MCT gates is not directly achievable in qubit-only circuits, hence they decompose their cyclic shift operator as shown in Fig. 1. Since there are $O(\log(N))$ parallel SWAP layers required for the implementation of the qubit permutation, the overall time complexity of cyclic shift operator is $O(\log(N - M + 1) \log(N))$.

Step 4: At this juncture, we look to see if the pattern string kept in the third register matches the cyclically moved text strings in the second register. Each of the first M bits in the second register and each of the M bits in the third register are combined using an XOR operation. For instance, the sequences match if the XOR outputs are all zeros. Then, with the use of CNOT gates on a quantum computer, we acquire,

$$\frac{1}{\sqrt{N-M+1}} \sum_{k=0}^{N-M} |k\rangle \, \text{CNOT}^{\otimes M} \left[\left(\bigotimes_{i=0}^{N-1} |t_{i+k}\rangle \right) \left(\bigotimes_{j=0}^{M-1} |p_j\rangle \right) \right]$$

$$= \frac{1}{\sqrt{N-M+1}} \sum_{k=0}^{N-M} \left[|k\rangle \left(\bigotimes_{i=0}^{N-1} |t_{i+k}\rangle \right) \left(\bigotimes_{j=0}^{M-1} |p_j \oplus t_{j+k}\rangle \right) \right].$$

For this purpose, the number of discrepancies between the pattern and the first M bits of the string register is stored in the final register. In fact, if and only if those two string parts match exactly, it is all zero.

Step 5: Finally, Grover's oracle which works on the pattern register is necessary to finish our algorithm because it will amplify and help in the identification of exact matches or near matches. We can get this oracle in $O(\log(M))$ depth using novel decomposition of MCT gate using intermediate qudits without ancilla qubits. For a better understanding, we have given an example of a proposed string-matching algorithm.

Example: Suppose we take an example string $\mathcal{T}$ (to be searched) = ABCDEFGH and Pattern, $\mathcal{P}$ (to be matched) = CDEFG. As per Fig. 6, $N = 8$ and $M = 5$. These M and N can be stated as $|t\rangle$ and $|p\rangle$ respectively. As per our proposed algorithm, we need an index register of $\lceil \log(N - M + 1) \rceil$ i.e., $\lceil \log(8 - 5 + 1) \rceil = 2$ extra qubits for cyclic shift operator as $|k\rangle$, which are initialized to 0. Next, we need $\frac{N}{2}$ ancilla qubits as $|a\rangle$ for parallel Fredkin operation, which is also initialized as 0. Finally one output qubit for Grover's search with 1 as input. So the initial quantum state is:

$$\psi_0 \to |k\rangle \otimes |t\rangle \otimes |p\rangle \otimes |a\rangle \otimes |o\rangle$$

$$\psi_0 \to |00\rangle \otimes |ABCDEFGH\rangle \otimes |CDEFG\rangle \otimes |0000\rangle \otimes |1\rangle$$

At first, we have to apply Hadamard transformation on the first two qubits, hence the quantum state evolves as,

$$\psi_1 \to \frac{1}{2} \begin{pmatrix} |00\rangle \otimes |ABCDEFGH\rangle \otimes |CDEFG\rangle \otimes |0000\rangle \otimes |1\rangle + \\ |01\rangle \otimes |ABCDEFGH\rangle \otimes |CDEFG\rangle \otimes |0000\rangle \otimes |1\rangle + \\ |10\rangle \otimes |ABCDEFGH\rangle \otimes |CDEFG\rangle \otimes |0000\rangle \otimes |1\rangle + \\ |11\rangle \otimes |ABCDEFGH\rangle \otimes |CDEFG\rangle \otimes |0000\rangle \otimes |1\rangle \end{pmatrix}$$

Now, the cyclic-shift operator comes into action. When the value of index register $|k\rangle$ is $|00\rangle$, there will be no change in the systems. For the value of $|01\rangle$, there will be one place cyclic shift of $|t\rangle$. For that, through MCT operations ancilla register $|a\rangle$ becomes $|1111\rangle$ first,

$$\psi_2 \to \frac{1}{2} \begin{pmatrix} |00\rangle \otimes |ABCDEFGH\rangle \otimes |CDEFG\rangle \otimes |0000\rangle \otimes |1\rangle + \\ |01\rangle \otimes |ABCDEFGH\rangle \otimes |CDEFG\rangle \otimes |1111\rangle \otimes |1\rangle + \\ |10\rangle \otimes |ABCDEFGH\rangle \otimes |CDEFG\rangle \otimes |0000\rangle \otimes |1\rangle + \\ |11\rangle \otimes |ABCDEFGH\rangle \otimes |CDEFG\rangle \otimes |0000\rangle \otimes |1\rangle \end{pmatrix}$$

We now perform parallel Fredkin operations to shift one place of the string $|t\rangle$ for the index register $|01\rangle$ as shown in Fig. 7,

$$\psi_3 \to \frac{1}{2} \begin{pmatrix} |00\rangle \otimes |ABCDEFGH\rangle \otimes |CDEFG\rangle \otimes |0000\rangle \otimes |1\rangle + \\ |01\rangle \otimes |BCDEFGHA\rangle \otimes |CDEFG\rangle \otimes |1111\rangle \otimes |1\rangle + \\ |10\rangle \otimes |ABCDEFGH\rangle \otimes |CDEFG\rangle \otimes |0000\rangle \otimes |1\rangle + \\ |11\rangle \otimes |ABCDEFGH\rangle \otimes |CDEFG\rangle \otimes |0000\rangle \otimes |1\rangle \end{pmatrix}$$

We now again get back the value of ancilla qubits to $|0000\rangle$ through inverse operations so that further cyclic-shift operations can be performed for other indexed values,

$$\psi_4 \to \frac{1}{2} \begin{pmatrix} |00\rangle \otimes |ABCDEFGH\rangle \otimes |CDEFG\rangle \otimes |0000\rangle \otimes |1\rangle + \\ |01\rangle \otimes |BCDEFGHA\rangle \otimes |CDEFG\rangle \otimes |0000\rangle \otimes |1\rangle + \\ |10\rangle \otimes |ABCDEFGH\rangle \otimes |CDEFG\rangle \otimes |0000\rangle \otimes |1\rangle + \\ |11\rangle \otimes |ABCDEFGH\rangle \otimes |CDEFG\rangle \otimes |0000\rangle \otimes |1\rangle \end{pmatrix}$$

Similarly, we perform a cyclic-shift operation for the other two index register's values, which are $|10\rangle$ and $|11\rangle$,

$$\psi_5 \to \frac{1}{2} \begin{pmatrix} |00\rangle \otimes |ABCDEFGH\rangle \otimes |CDEFG\rangle \otimes |0000\rangle \otimes |1\rangle + \\ |01\rangle \otimes |BCDEFGHA\rangle \otimes |CDEFG\rangle \otimes |0000\rangle \otimes |1\rangle + \\ |10\rangle \otimes |CDEFGHAB\rangle \otimes |CDEFG\rangle \otimes |0000\rangle \otimes |1\rangle + \\ |11\rangle \otimes |DEFGHABC\rangle \otimes |CDEFG\rangle \otimes |0000\rangle \otimes |1\rangle \end{pmatrix}$$

At this point, we perform the XOR operation with the use of CNOT gates between the first M bits of $|t\rangle$ register and M bits of $|p\rangle$ register. Outputs of $|p\rangle$ are all zeros for the indexed value of $|10\rangle$,

$$\psi_6 \to \frac{1}{2} \begin{pmatrix} |00\rangle \otimes |ABCDEFGH\rangle \otimes |CDEFG\rangle \otimes |0000\rangle \otimes |1\rangle + \\ |01\rangle \otimes |BCDEFGHA\rangle \otimes |CDEFG\rangle \otimes |0000\rangle \otimes |1\rangle + \\ |10\rangle \otimes |CDEFGHAB\rangle \otimes |00000\rangle \otimes |0000\rangle \otimes |1\rangle + \\ |11\rangle \otimes |DEFGHABC\rangle \otimes |CDEFG\rangle \otimes |0000\rangle \otimes |1\rangle \end{pmatrix}$$

We now perform the bit-flip operation through X gate on $|p\rangle$ to get all ones,

$$\psi_7 \to \frac{1}{2} \begin{pmatrix} |00\rangle \otimes |ABCDEFGH\rangle \otimes |CDEFG\rangle \otimes |0000\rangle \otimes |1\rangle + \\ |01\rangle \otimes |BCDEFGHA\rangle \otimes |CDEFG\rangle \otimes |0000\rangle \otimes |1\rangle + \\ |10\rangle \otimes |CDEFGHAB\rangle \otimes |11111\rangle \otimes |0000\rangle \otimes |1\rangle + \\ |11\rangle \otimes |DEFGHABC\rangle \otimes |CDEFG\rangle \otimes |0000\rangle \otimes |1\rangle \end{pmatrix}$$

Next we perform Hadamard operation on output qubit $|o\rangle$ to perform the Grover's search,

$$\psi_8 \to \frac{1}{2\sqrt{2}} \begin{bmatrix} |00\rangle \otimes |ABCDEFGH\rangle \otimes |CDEFG\rangle \otimes |0000\rangle \otimes |0\rangle + \\ |01\rangle \otimes |BCDEFGHA\rangle \otimes |CDEFG\rangle \otimes |0000\rangle \otimes |0\rangle + \\ |10\rangle \otimes |CDEFGHAB\rangle \otimes |11111\rangle \otimes |0000\rangle \otimes |0\rangle + \\ |11\rangle \otimes |DEFGHABC\rangle \otimes |CDEFG\rangle \otimes |0000\rangle \otimes |0\rangle - \\ (|00\rangle \otimes |ABCDEFGH\rangle \otimes |CDEFG\rangle \otimes |0000\rangle \otimes |1\rangle + \\ |01\rangle \otimes |BCDEFGHA\rangle \otimes |CDEFG\rangle \otimes |0000\rangle \otimes |1\rangle + \\ |10\rangle \otimes |CDEFGHAB\rangle \otimes |11111\rangle \otimes |0000\rangle \otimes |1\rangle + \\ |11\rangle \otimes |DEFGHABC\rangle \otimes |CDEFG\rangle \otimes |0000\rangle \otimes |1\rangle) \end{bmatrix}$$

We now perform the MCT operation between $|p\rangle$ and $|o\rangle$ and quantum state evolves as,

$$\psi_9 \rightarrow \frac{1}{2\sqrt{2}} \begin{bmatrix} |00\rangle \otimes |ABCDEFGH\rangle \otimes |CDEFG\rangle \otimes |0000\rangle \otimes |0\rangle + \\ |01\rangle \otimes |BCDEFGHA\rangle \otimes |CDEFG\rangle \otimes |0000\rangle \otimes |0\rangle + \\ |10\rangle \otimes |CDEFGHAB\rangle \otimes |11111\rangle \otimes |0000\rangle \otimes |1\rangle + \\ |11\rangle \otimes |DEFGHABC\rangle \otimes |CDEFG\rangle \otimes |0000\rangle \otimes |0\rangle - \\ (|00\rangle \otimes |ABCDEFGH\rangle \otimes |CDEFG\rangle \otimes |0000\rangle \otimes |1\rangle + \\ |01\rangle \otimes |BCDEFGHA\rangle \otimes |CDEFG\rangle \otimes |0000\rangle \otimes |1\rangle + \\ |10\rangle \otimes |CDEFGHAB\rangle \otimes |11111\rangle \otimes |0000\rangle \otimes |0\rangle + \\ |11\rangle \otimes |DEFGHABC\rangle \otimes |CDEFG\rangle \otimes |0000\rangle \otimes |1\rangle) \end{bmatrix}$$

We next perform the mirror operations to get back the quantum register $|p\rangle$ to its initial state,

$$\psi_{10} \rightarrow \frac{1}{2\sqrt{2}} \begin{bmatrix} |00\rangle \otimes |ABCDEFGH\rangle \otimes |CDEFG\rangle \otimes |0000\rangle \otimes |0\rangle + \\ |01\rangle \otimes |BCDEFGHA\rangle \otimes |CDEFG\rangle \otimes |0000\rangle \otimes |0\rangle + \\ |10\rangle \otimes |CDEFGHAB\rangle \otimes |CDEFG\rangle \otimes |0000\rangle \otimes |1\rangle + \\ |11\rangle \otimes |DEFGHABC\rangle \otimes |CDEFG\rangle \otimes |0000\rangle \otimes |0\rangle - \\ (|00\rangle \otimes |ABCDEFGH\rangle \otimes |CDEFG\rangle \otimes |0000\rangle \otimes |1\rangle + \\ |01\rangle \otimes |BCDEFGHA\rangle \otimes |CDEFG\rangle \otimes |0000\rangle \otimes |1\rangle + \\ |10\rangle \otimes |CDEFGHAB\rangle \otimes |CDEFG\rangle \otimes |0000\rangle \otimes |0\rangle + \\ |11\rangle \otimes |DEFGHABC\rangle \otimes |CDEFG\rangle \otimes |0000\rangle \otimes |1\rangle) \end{bmatrix}$$

Finally, we perform Grover's amplitude amplification to obtain the final outcome, $\psi_{11} \rightarrow -(|10\rangle \otimes |ABCDEFGH\rangle \otimes |CDEFG\rangle \otimes |0000\rangle \otimes |1\rangle)$

The final quantum state depicts that the pattern '$CDEFG$' has a match in string '$ABCDEFG$'. The pattern can be found from the third position in the string. We can easily get the pattern's position by adding two with the first position of the string since the indexed value suggests us as 10 in binary i.e., two in integer. We also verify our results through simulation on the QuDiet platform [ ].

4 Discussion

4.1 Improved Time Complexity

We determine our algorithm's time complexity in this subsection. Strings $\mathcal{T}$ and $\mathcal{P}$ require $O(1)$ time to encode. It also requires $O(1)$ time to apply the Hadamard or Fourier transform to the index register. The time required by the cyclic-shift operator $\mathcal{S}$ is $O((\log(N-M+1)\log(N))$. It takes time $O(1)$ to evaluate XOR outcomes using CNOT gates because they allow for simple concurrent processing.

Last but not least, the complexity of the Grover oracle is $O(\log(M))$. A single Grover step's complexity, which accounts for all the steps taken into account so far, is then $O\left(\log(N - M + 1)\log(N) + \log(M)\right)$. The Grover steps must be repeated $O(\sqrt{N - M + 1})$ times in order for the Grover search to be successful. With this added intricacy, the total complexity is now:
$O\left(\sqrt{N - M + 1}\left((\log(N - M + 1)\log(N)) + \log(M)\right)\right).$

4.2 Improved Space Complexity

We also need $O(\log(N - M + 1))$ qubits for the index register in addition to the N and M qubits required to store the search string and the pattern. We require $\frac{N}{2}$ ancilla qubits for the index register in order to implement our cyclic-shift operator in a depth-optimized manner. We do not need any other extra ancilla qubit for our proposed approach. A comparative study of space complexity with [ ] is exhibited in Table 2.

Space complexity	[ ]	This work
Data qubit	$N + M + \log N$	$N + M + \log(\lceil N - M + 1 \rceil)$
Ancilla qubit	$\frac{N}{2} + M$	$\frac{N}{2}$

Table 2: Comparison of space complexity of our work with state-of-the-art algorithm [ ].

4.3 Improved Circuit Cost

One can calculate the gate count in terms of CNOT and T gates according to the state-of-the-art circuit. Since it is widely anticipated that T gates will predominate the cost of implementation in the fault-tolerant regime, assuming the standard gate set of Clifford+T, they chose those two gates as metrics. The best part of our proposed algorithm is no T gate is required for string-matching. We estimate the gate count in terms of CNOT, ternary CNOT and quaternary CNOT gates. It is important to remember the fact that the n-qubit Toffoli decomposition has a logarithmic depth and uses a maximum of $n + 1$ ternary CNOT gates and $n - 4$ quaternary CNOT gates. A comparative study of circuit cost with [ ] is exhibited in Table 3.

The cost of the encoding step is zero because the strings $\mathcal{T}$ and $\mathcal{P}$ can be originally encoded in qubits in the $|0\rangle$ state using only the identity and bit-flip(X) gates. Hadamard gates are required for a Hadamard transformation of the index register which also requires zero cost. The stated permutation of size as large as

Circuit cost	[ ]	This work
T	$(8M - 17 + 7(N-1)O(\log(N))) \times 2\sqrt{N}$	0
CNOT	$(7M - 12 + (8N - 9)O(\log(N))) \times 2\sqrt{N}$	$((2N-1)O(\log(N)) + M) \times 2\sqrt{N-M+1}$
ternary CNOT	0	$((N - M + 2)O(\log(N - M + 1)) + (3N - 1)O(\log(N)) + (M - 1)\log(M)) \times 2\sqrt{N-M+1}$
quaternary CNOT	0	$((N - M - 3)O(\log(N - M + 1)) + (M - 4)\log(M)) \times 2\sqrt{N-M+1} \times 2\sqrt{N-M+1}$

Table 3: Comparison of circuit cost of our work with state-of-the-art algorithm [ ].

N can be divided into at most $N - 1$ transpositions, so the cyclic shift operator $\mathcal{S}$ consists of an MCT gate with depth $O(\log(N - M + 1))$ and at most $N - 1$ Fredkin gates. As per our proposed Fredkin gate, each Fredkin gate costs 2 CNOT gates and 3 ternary CNOT gates. Thus the cyclic shift operator costs at most $(2N - 1)O(\log(N))$ CNOT gates, and $((N - M + 2)O(\log(N - M + 1)) + (3N - 1)O(\log(N))$ ternary CNOT gates, and $((N - M - 3)O(\log(N - M + 1))$ quaternary CNOT gates. Next, the XOR operation requires M CNOT gates. Lastly, the Grover oracle with multi-controlled Toffoli decomposition with intermediate qudits can be implemented with $(M + 1)\log(M)$ ternary CNOT gates and $(M - 4)\log M$ quaternary CNOT gates without any ancilla. Lastly, for amplitude amplification, we need to repeat this $\sqrt{N - M + 1}$ times. The total CNOT, ternary CNOT, and quaternary CNOT count are, thus, given in Table 3 where the component of 2 comes from the necessity of applying a unitary to create the states $|\psi\rangle = U|0\rangle$ and the inverse unitary $U^\dagger$ in order to amplify the amplitude.

4.4 Error Analysis

Any quantum system is susceptible to different types of errors such as decoherence, and noisy gates. For a binary quantum system, the gate error scales as 2^2 and 2^4 for 1- and 2-qubit gates respectively [ ]. Furthermore, for qubits, the amplitude damping error decays the state $|1\rangle$ to $|0\rangle$ with probability λ_1. For a higher dimensional system, every state in level $|i\rangle \neq |0\rangle$ has a probability λ_1 of decaying. In other words, the usage of higher dimensional states penalizes the system with more errors. Nevertheless, the effect of these errors [ ] on the used decomposition of the Multi-controlled Toffoli gate has been studied by Saha et al. [ ] and the used decomposition of Toffoli gate for Fredkin gate has been studied by Gokhale et. al. [ ]. They have demonstrated that even though the use of intermediate qudits results in a rise in error, the total error probability of the decomposition is lower than the ones used currently because there are fewer gates and less depth [ ]. This interpretation is also applicable to our approach of solving the string-matching problem, since the gate cost and the depth have been reduced as compared to [ ]. Hence, we claim that our solution for the string-matching problems with intermediate qudits is superior in terms of error efficiency as compared to [ ]. The generalized Toffoli decomposition of [ ] that has been used in our proposed circuits for string-matching is also efficient with respect to crosstalk errors [ ] due to its crosstalk-aware structure. Since the Fredkin gate decomposition is new to this paper, we show the probability of success for the Fredkin gate decomposition using the method of [ ] and our proposed method. As shown in Fig. 8, we find that the decomposition in [ ] has a considerably higher error rate than the one we propose. This is explained by our decomposition's shallower depth and fewer gates. The advantage of our decomposition lies in the general substantial reduction in the gate count and the depth, despite the fact that some ternary and quaternary gates are used, which have a greater error probability due to the plague of dimensionality. Thus we can conclude that our circuit for string-matching is relevant instead of using higher dimensional qudits through error analysis.

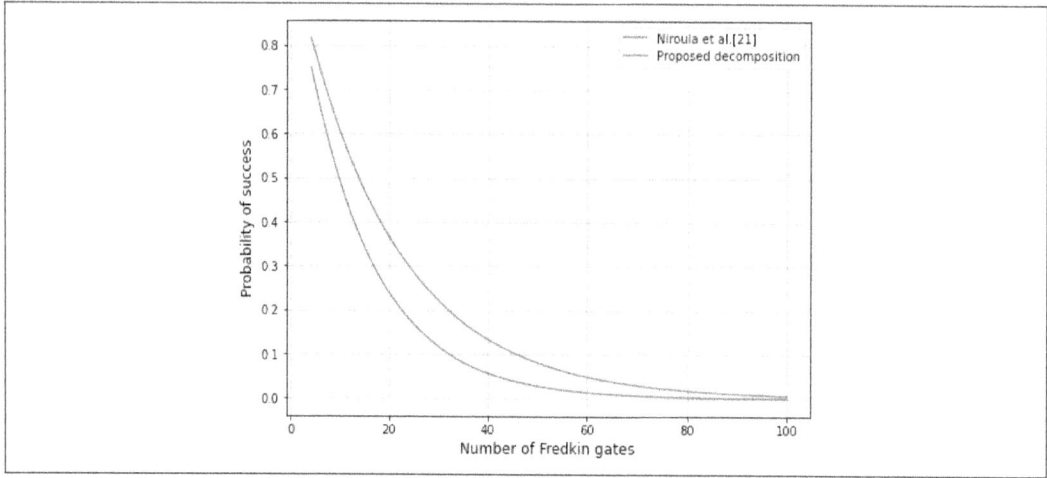

Figure 8: Probability of success for the decomposition of Fredkin gate using our proposed method (upper curve) versus the method in [ ] (lower curve).

5 Conclusion

We have built a quantum string-matching algorithm in this work that allows for a circuit-depth complexity of $O\left(\sqrt{N-M+1}\left((\log(N-M+1)\log(N)) + \log(M)\right)\right)$. Additionally, we offer a detailed gate-level version of our method, allowing for a precise calculation of the required quantum resources. This circuit for string-matching can be designed for any dimensional quantum system since the used gates are generalized, which makes the proposed approach generalized in nature. The primary use cases of the matching algorithm, such as a quick text search in a big file or spotting patterns in an image, can now be carried out more effectively. The proposed decomposition of the Fredkin gate can be used in other algorithms for their efficient implementation. In fact, the overall findings show great promise for future work on effectively implementing other algorithms in intermediate qudit-assisted quantum computing. Whether the proposed approach will perform with similar efficiency in fault-tolerant regime [ ] can only be answered with the evolution of more scalable qudit-supported quantum hardware. Thus it is kept as a future aspect of this work when error correction for qutrits and ququarts would be feasible.

Acknowledgments

There is no conflict of interest.

References

[1] Matthew Amy, Dmitri Maslov, Michele Mosca, and Martin Roetteler. A meet-in-the-middle algorithm for fast synthesis of depth-optimal quantum circuits. *IEEE Transactions on Computer-Aided Design of Integrated Circuits and Systems*, 32(6):818–830, 2013.

[2] A. Barenco, C. H. Bennett, R. Cleve, D. P. DiVincenzo, N. Margolus, P. Shor, T. Sleator, J. A. Smolin, and H. Weinfurter. Elementary gates for quantum computation. *Physical Review A*, 52:3457–3467, 1995.

[3] Turbasu Chatterjee, Arnav Das, Subhayu Kumar Bala, Amit Saha, Anupam Chattopadhyay, and Amlan Chakrabarti. Qudiet: A classical simulation platform for qubit-qudit hybrid quantum systems. *IET Quantum Communication*, page 167–180, 2023.

[4] Yao-Min Di and Hai-Rui Wei. Synthesis of multivalued quantum logic circuits by elementary gates. *Phys. Rev. A*, 87:012325, 2013.

[5] Mark Ettinger and Peter Hoyer. A quantum observable for the graph isomorphism problem. *arXiv:quant-ph/9901029*, 1999.

[6] Laurin E. Fischer, Daniel Miller, Francesco Tacchino, Panagiotis Kl. Barkoutsos, Daniel J. Egger, and Ivano Tavernelli. Ancilla-free implementation of generalized measurements for qubits embedded in a qudit space. *Phys. Rev. Res.*, 4:033027, 2022.

[7] Pranav Gokhale, Jonathan M. Baker, Casey Duckering, Natalie C. Brown, Kenneth R. Brown, and Frederic T. Chong. Asymptotic improvements to quantum circuits via qutrits. In *Proceedings of the 46th International Symposium on Computer Architecture*, ISCA '19, page 554–566, New York, NY, USA, 2019. Association for Computing Machinery.

[8] L. K. Grover. A fast quantum mechanical algorithm for database search. In *Proceedings of the Twenty-eighth Annual ACM Symposium on Theory of Computing*, STOC '96, pages 212–219, New York, NY, USA, 1996. ACM.

[9] Ramesh Hariharan and V. Vinay. String matching in Õ(sqrt(n)+sqrt(m)) quantum time. *J. Discrete Algorithms*, 1:103–110, 2003.

[10] Donald E. Knuth, James H. Morris, Jr., and Vaughan R. Pratt. Fast pattern matching in strings. *SIAM Journal on Computing*, 6(2):323–350, 1977.

[11] Ritajit Majumdar, Dhiraj Madan, Debasmita Bhoumik, Dhinakaran Vinayagamurthy, Shesha Raghunathan, and Susmita Sur-Kolay. Optimizing ansatz design in qaoa for max-cut. *arXiv preprint arXiv:2106.02812*, 2021.

[12] Ritajit Majumdar, Amit Saha, Amlan Chakrabarti, and Susmita Sur-Kolay. Intermediate qutrit-assisted Toffoli gate decomposition with quantum error correction. *Quantum Inf Process*, 23, 2024.

[13] Ashley Montanaro. Quantum pattern matching fast on average. *Algorithmica*, 77(1):16–39, 2017.

[14] Prakash Murali, David C. Mckay, Margaret Martonosi, and Ali Javadi-Abhari. Software mitigation of crosstalk on noisy intermediate-scale quantum computers. In *Proceedings of the Twenty-Fifth International Conference on Architectural Support for Programming Languages and Operating Systems*, ASPLOS '20, page 1001–1016, New York, NY, USA, 2020. Association for Computing Machinery.

[15] Ashok Muthukrishnan and C. R. Stroud. Multivalued logic gates for quantum computation. *Phys. Rev. A*, 62:052309, 2000.

[16] Gonzalo Navarro. A guided tour to approximate string matching. *ACM Comput. Surv.*, 33(1):31–88, 2001.

[17] M. A. Nielsen and I. L. Chuang. *Quantum Computation and Quantum Information: 10th Anniversary Edition*. Cambridge University Press, 2010.

[18] A. S. Nikolaeva, E. O. Kiktenko, and A. K. Fedorov. Efficient realization of quantum algorithms with qudits. *arXiv:2111.04384*, 2021.

[19] A. S. Nikolaeva, E. O. Kiktenko, and A. K. Fedorov. Decomposing the generalized Toffoli gate with qutrits. *Physical Review A*, 105(3), 2022.

[20] Anstasiia S. Nikolaeva, Evgeniy O. Kiktenko, and Aleksey K. Fedorov. Generalized Toffoli Gate Decomposition Using Ququints: Towards Realizing Grover's Algorithm with Qudits. *Entropy*, 25(2):387, 2023.

[21] Pradeep Niroula and Yunseong Nam. A quantum algorithm for string matching. *npj Quantum Information*, 7:37, 2021.

[22] John Preskill. Quantum Computing in the NISQ era and beyond. *Quantum*, 2:79, 2018.

[23] Amit Saha, Ritajit Majumdar, Debasri Saha, Amlan Chakrabarti, and Susmita Sur-Kolay. Asymptotically improved circuit for a d-ary Grover's algorithm with advanced decomposition of the n-qudit Toffoli gate. *Phys. Rev. A*, 105:062453, 2022.

[24] Yuchen Wang, Zixuan Hu, Barry C. Sanders, and Sabre Kais. Qudits and high-dimensional quantum computing. *Frontiers in Physics*, 8, 2020.

SOME CONSISTENCY RESULTS FOR MANY-VALUED JUDGMENT AGGREGATION

CHRISTIAN G. FERMÜLLER
Institute of Logic & Computation, TU Wien, 1040 Vienna, Austria
chrisf@logic.at

SEBASTIAN UHL
Institute of Logic & Computation, TU Wien, 1040 Vienna, Austria
sebastian.uhl22@gmail.com

Abstract

Judgment aggregation (JA) poses the problem of finding a consistent collective judgment for a set of logically related propositions based on judgments of individuals. There are well-known impossibility results for classical JA, which have recently been extended to non-classical logics, including many-valued logics. We complement these negative results with some positive results. We first study average aggregation, which is arguably the most natural rule in a many-valued setting, and show how to generate consistent aggregated judgments in either Kleene-Zadeh or Łukasiewicz logic for certain types of agendas. We then generalize these results to a wider class of aggregation rules applied to judgments using Kleene-Zadeh and Gödel logic by imposing a restricted systematicity condition. Finally, we introduce median aggregation and show a possibility result that applies to arbitrary many-valued logics by generalizing List's profile condition of unidimensional alignment to a many-valued setting.

1 Introduction

Judgment aggregation (JA) is a relatively recent research topic in the intersection of economy, logic, computer science, and mathematics, closely related to social choice and voting theory (see, e.g., [9, 15]). Classical JA is concerned with the aggregation of individual opinions (judgments) on statements (elements of an agenda) into a collective judgment. The agenda is represented as a set of propositional formulas, each

---

The paper is based on results introduced by the first author at ISMVL 2023 [8] and, jointly with the second author, at IUKM 2023 [21].

of which is either to be rejected or accepted. Reasonable aggregation rules should deliver consistent collective judgments and satisfy certain rationality constraints. In particular, the aggregation rule should not be dictatorial; i.e., it should not ignore the opinions of all but one of the agents. Impossibility results state that there are no reasonable, non-dictatorial aggregation rules that yield a consistent aggregation of arbitrary consistent individual judgments for a large class of agendas (see, e.g., [4, 6]). These results have been generalized to non-classical logics, in particular also to a many-valued setting, where the judgments are modeled as valuations of propositional formulas according to some given many-valued logic (see, e.g., [11, 7]). This situation motivates the search for criteria guaranteeing that natural, non-dictatorial aggregation rules yield consistent collective judgments for at least certain types of agendas. Classical JA results of the latter kind are summarized in [9].

Motivated by an analysis of the so-called doctrinal paradox [12] in Section 2 and the corresponding discursive dilemma [20] from a many-valued point of view in Section 3, we first focus on average aggregation in this paper. This is arguably the most natural way of aggregating individual judgments using the unit interval $[0, 1]$ instead of just $\{0, 1\}$ as set of truth values. Two specific many-valued logics emerge naturally from our take on the discursive dilemma: Kleene-Zadeh logic KZ, where conjunction and disjunction are interpreted by $\min(x, y)$ and $\max(x, y)$, respectively; and Łukasiewicz logic Ł, where $\max(0, x+y-1)$ interprets conjunction and $\min(1, x+y)$ interprets disjunction. In both cases, the truth function for negation is $1-x$. In Section 4, we mainly consider a scenario where the individual judgments of the agents are Boolean but where these judgments are aggregated by averaging and hence are many-valued. We identify conditions for agendas as well as for judgment profiles that guarantee that averaging yields either KZ-consistent or Ł-consistent collective judgments. For the case of KZ, we will generalize this scenario to one where not only the collective but also the individual judgments of the agents are many-valued (see Section 4.2). We do not confine ourselves to average aggregation but rather identify in Section 5 a systematicity condition for aggregation functions that support consistent collective judgments. Moreover, in Section 6 we will show a similar possibility result for Gödel(-Dummett) logic G. Finally, in Section 7 we introduce so-called median aggregation. This turns out to be a rather flexible form of aggregation that fits JA scenarios based on arbitrary many-valued logics. We show that a many-valued version of List's profile condition of unidimensional alignment [14] guarantees consistent collective judgments for any many-valued logic.

2 Classical judgment aggregation

2.1 The doctrinal paradox and discursive dilemma

Challenges for judgment aggregation are often motivated by the so-called *doctrinal paradox*, referring to the following example, due to the legal scholars Kornhauser and Sager [12]. (Our presentation follows [17].) A plaintiff has brought a civil suit against a defendant, alleging a breach of contract between them. Three judges (J_1, J_2, J_3) have to determine whether the defendant must pay damages to the plaintiff (d or $\neg d$). The case brings up two issues, namely, whether the contract was valid (v or $\neg v$), and whether the defendant was in breach of it (b or $\neg b$). Contract law stipulates that the defendant must pay damages if and only if the contract was valid and he was in breach of it ($d \leftrightarrow v \wedge b$). Suppose that the judges have the following views on the two issues, and, accordingly, of the case:

J_1:	v	$\neg b$	$\neg d$
J_2:	$\neg v$	b	$\neg d$
J_3:	v	b	d

The 'paradox' arises from the fact that two contradictory legal doctrines might apply here.

(1) Since the majority judges $\neg d$, no damage is to be paid.

(2) Since each of the two conditions, v and b, are judged to hold by a majority, the defendant must pay damages.

Pettit [20] introduced a general version of the above scenario, dubbed *discursive dilemma*. His presentation shifts the focus away from the conflict between the two methods of finding an adequate overall judgment and instead makes the logical inconsistency of the set of propositions that are accepted by a majority of the judges explicit. In the above example, this inconsistent set is $\{v, b, d \leftrightarrow v \wedge b, \neg d\}$.

2.2 The classical JA setting

As presented in [20], for general classical judgment aggregation (JA) one stipulates an *agenda* consisting of a set of positive (unnegated) formulas (*issue set*) and their negations. Each *agent (judge)* selects a consistent and complete subset of the agenda, representing their individual *judgments*. Here, consistency refers to classical logic, and *completeness* means that for each ϕ in the issue set, either ϕ or $\neg \phi$ is present. These judgments are to be *aggregated* into a single subset of the agenda, intended to reflect the group decision regarding the agenda. List and Pettit [16] completed the model by explicitly referring to *judgment aggregation functions* in this context and formulating various minimal conditions for them.

We fix some corresponding formal notions, following [9].

Definition 1. *Let $\mathcal{L}$ be a propositional language (set of formulas) over a given set of atoms At. A (classical) judgment aggregation problem (JA problem) over $\mathcal{L}$ is a pair $\langle N, \mathcal{A} \rangle$, where*

- $N = \{1, \ldots, n\}$, where $n = |N| \geq 1$, is the set of agents (judges);
- $\mathcal{A} = I \cup \{\neg \phi \colon \phi \in I\}$, where the finite issue set $I \subseteq \mathcal{L}$ contains only positive formulas. $\mathcal{A}$ is called agenda.

Remark 1. *It suffices to restrict our attention to formulas involving conjunction ($\wedge$), disjunction ($\vee$), and negation ($\neg$). Other logical connectives, like implication ($\rightarrow$) and bi-implication ($\leftrightarrow$), can be defined from the former ones as usual.*

Definition 2. *Given a JA problem $\langle N, \mathcal{A} \rangle$, a judgment (set) J is a consistent and complete subset of $\mathcal{A}$. The set of all judgment sets is denoted by $\mathbf{J}$. A judgment profile is an n-tuple $\langle J_i \rangle_{i \in N}$ of judgment sets from $\mathbf{J}^N$. For a formula $\phi \in \mathcal{A}$, $P_\phi = \{i \in N \colon \phi \in J_i\}$ denotes the set of agents accepting ϕ. $\mathbf{P}$ denotes the set of all judgment profiles.*

Definition 3. *Given a JA problem $\langle N, \mathcal{A} \rangle$ an* aggregation rule *(or* aggregation function*) for it is of type $F \colon \mathbf{P} \to 2^\mathcal{A}$ mapping each judgment profile $P = \langle J_i \rangle_{i \in N}$ into a subset $F(P)$ of $\mathcal{A}$, called a* collective set *or* collective judgment.

Aggregation rules are total functions. This means that there is a corresponding collective set for every judgment profile. This is commonly referred to as *universal domain condition*.

Note that the collective set is not consistent nor complete, in general. That is, the collective set need not be a judgment set.

We refer to [9] and [13] for an overview of various types of aggregation rules. Important examples are the *majority rule* F_{maj}, the *unanimity rule* F_u, and the *quota rule* F_t, where $t = \langle t_\phi \rangle_{\phi \in \mathcal{A}} \in \{0, \ldots, n\}^\mathcal{A}$ is a tuple of integers or quotas, one for each formula in the agenda. These rules are defined for $P \in \mathbf{P}$ respectively as follows:

- $F_{\mathrm{maj}}(P) = \{\phi \in \mathcal{A} \colon |P_\phi| \geq \lceil (n+1)/2 \rceil\} = \{\phi \in \mathcal{A} \colon |P_\phi| > n/2\}$,
- $F_\mathrm{u}(P) = \{\phi \in \mathcal{A} \colon |P_\phi| = n\}$,

- $F_t(P) = \{\phi \in \mathcal{A} \colon |P_\phi| \geq t_\phi\}$.

In the last case, a formula ϕ is collectively accepted iff at least t_ϕ agents accept it. Clearly, F_{maj} and F_{u} are instances of the more general quota rule.

2.3 A classical impossibility result

Using the terminology introduced above, we may rephrase the discursive dilemma as follows. The majority rule F_{maj} does not always yield a *consistent* collective set—and hence no judgment (consistent valuation)—for the issue set $I = \{v, b, d, v \wedge b \leftrightarrow d\}$. Actually, the statement already holds for $I = \{v, b, v \wedge b\}$. Moreover, it also holds for F_{u} and for every instance of F_t. Of course, it is trivial to define aggregation functions that always deliver consistent sets. But it seems that all such functions disrespect the judgments of some agents if the agenda is not trivial, i.e., if some issues are logically dependent on others. These observations have been generalized by so-called impossibility results. In order to formulate a representative example of an impossibility theorem precisely, we have to define agenda conditions. We remind the reader that the following conditions refer to classical logic and that a set of formulas is *minimally inconsistent* if it is inconsistent, while every proper subset of it is consistent.

Definition 4. *An agenda $\mathcal{A}$ is* non-simple *if it contains a minimally inconsistent subset $X \subseteq \mathcal{A}$, such that $|X| \geq 3$.*

Definition 5. *An agenda $\mathcal{A}$ is* evenly negatable *if it contains a minimally inconsistent subset $X \subseteq \mathcal{A}$ and a subset $Y = \{\phi, \psi\} \subseteq \mathcal{A}$ such that $(X - Y) \cup \{\neg\phi, \neg\psi\}$ is consistent.*

We specify a few properties of possible aggregation rules.

Definition 6. *Given a JA problem $\langle N, \mathcal{A} \rangle$ and the corresponding set $\mathbf{P}$ of judgment profiles, an aggregation rule $F \colon \mathbf{P} \to 2^{\mathcal{A}}$ is*

- rational *iff $\forall P \in \mathbf{P} \colon F(P)$ is consistent and complete;*
- anonymous *iff $\forall P, P' \in \mathbf{P}$ s.t. P' is a permutation of $P \colon F(P) = F(P')$;*
- unanimous *iff $\forall \phi \in \mathcal{A}, \forall P = \langle J_i \rangle_{i \in N} \in \mathbf{P} \colon [\forall i \in N\ \phi \in J_i]$ implies $\phi \in F(P)$;*
- systematic *iff $\forall \phi, \psi \in \mathcal{A}, \forall P = \langle J_i \rangle_{i \in N}, P' = \langle J'_i \rangle_{i \in N} \in \mathbf{P} \colon$ if $[\forall i \in N\ \phi \in J_i$ iff $\psi \in J'_i]$, then $[\phi \in F(P)$ iff $\psi \in F(P')]$;*
- dictatorial *iff $\exists i \in N$ s.t. $\forall P = \langle J_i \rangle_{i \in N} \in \mathbf{P} \colon F(P) = J_i$.*

Note that the aggregation rules F_{maj} and F_{u} mentioned in 2.2, are anonymous, unanimous and systematic. The quota rule F_t is anonymous and unanimous, but only systematic if certain uniformity conditions on the thresholds associated with the agenda items are satisfied. Arguably, these three conditions are *desiderata* for any reasonable form of judgment aggregation. On the other hand, we want to avoid dictatorial aggregation. The following theorem shows that this is impossible.

Theorem 1 ([4]). *If the agenda $\mathcal{A}$ of a JA problem $\langle N, \mathcal{A} \rangle$ is non-simple and evenly negatable, then every aggregation rule F is rational, unanimous and systematic iff it is dictatorial.*

3 Many-valued judgment aggregation

3.1 Two levels of applying many-valued logics

We will consider two scenarios involving many-valued logic for the aggregation of judgments.

(1) Sticking with Boolean judgments on propositions by the individual agents, one may employ many-valued logics for aggregating those classical (Boolean) judgments.

(2) The agents may be asked to assign truth values extending the classical truth value set $\{0, 1\}$; aggregation of such judgments is then naturally many-valued as well.

Note that (1) is a special case of (2). We claim that it is nevertheless useful to treat these levels of many-valued judgment aggregation separately, since in many situations one has only Boolean judgments of agents to work with. For example, one may deem it too demanding for the agents to agree on a particular many-valued logic and then to remain consistent in all their judgments with respect to the chosen many-valued logic. In such situations, it is more appropriate to stick with option (1) and ask the agents for unqualified rejection or acceptance of propositions. On the other hand, we are all familiar with questionnaires, which ask individuals to specify their *degree* of acceptance of given propositions on a given scale. Consistently summarizing corresponding replies in the form of *most representative answers* may be considered a judgment aggregation problem of type (2).[1]

---

[1] The latter scenario is sometimes called *attitude aggregation* (see, e.g., [11]). Here we prefer to stick with the more familiar term 'judgment aggregation' also when the agents assign intermediate truth values.

To clarify the appropriate notion of consistency for the many-valued setting we need to fix some notions.

Definition 7. *A (general) propositional language $\mathcal{L}$ is the set of formulas built up from atoms (propositional variables) in At using connectives in $Op_\mathcal{L}$. More precisely:*

- *$At \subseteq \mathcal{L}$,*

- *if $\phi_1, \ldots, \phi_n \in \mathcal{L}$, then $\circ(\phi_1, \ldots, \phi_n) \in \mathcal{L}$ for every n-ary $\circ \in Op_\mathcal{L}$.*[2]

A many-valued logic Λ over the set of truth values $[0, 1]$ is specified by associating with each n-ary $\circ \in Op_\mathcal{L}$ a corresponding truth function $\tilde{\circ} : [0,1]^n \to [0,1]$. *A Λ-valuation V of a set of formulas $X \subseteq \mathcal{L}$ is an assignment of truth values to the formulas in X that respects the truth functions associated with Λ, i.e., a Λ-valuation is a homomorphic extension of some value assignment to the atoms. An arbitrary assignment of values in $[0,1]$ to formulas in X is called Λ-consistent if it is a Λ-valuation; otherwise, it is called Λ-inconsistent.*

For the rest of the paper we restrict attention to the language containing only the connectives $\wedge$, $\vee$, $\neg$ and $\to$. This is motivated by scenario (1), above, where classical formulas are interpreted via many-valued truth functions in the aggregation procedure.

In the classical setting (Section 2.2), agendas have been defined as sets of positive formulas, plus their negations and judgments have been represented as consistent and complete subsets of agendas. In the many-valued setting, it is more appropriate to identify judgments with valuations of the issue set.

Definition 8. *As in the classical case, also in the many-valued setting, a* judgment aggregation problem *(JA problem) is a pair $\langle N, \mathcal{A} \rangle$, where*

- *$N = \{1, \ldots, n\}$, where $n = |N| \geq 1$, is the set of* agents (judges);

- *the* agenda *$\mathcal{A}$ is any finite, non-empty set of formulas.*

The agenda $\mathcal{A}$ is called closed *if for each $\phi \in \mathcal{A}$ the subformulas of ϕ occur in $\mathcal{A}$. Given a many-valued logic Λ, a Λ-judgment $J : \mathcal{A} \to [0,1]$ is a Λ-valuation of $\mathcal{A}$. The set of all Λ-judgments is denoted by $\mathbf{J}_\Lambda$. A corresponding* judgment profile *$P = \langle J_i \rangle_{i \in N} \in \mathbf{J}_\Lambda^N$ over Λ is an n-tuple of Λ-judgments, where J_i denotes the judgment of agent $i \in N$. $\mathbf{P}_\Lambda$ denotes the set of all corresponding judgment profiles over Λ.*

---

[2] As usual, we will use infix notation for binary connectives and drop parentheses whenever no ambiguity arises in the given context.

Definition 9. *A many-valued aggregation rule (or aggregation function) for a given JA problem $\langle N, \mathcal{A}\rangle$ is of type $F : \mathbf{P}_\Lambda \to [0,1]^\mathcal{A}$ for some logic Λ. F maps any given Λ-judgment profile $P \in \mathbf{P}_\Lambda$ into the collective (aggregated) judgment $F(P)$. If $F(P)$ is a Λ-valuation, then the collective judgment is called Λ-consistent.*

Remark 2. *Note that an aggregated judgment may well be Λ-inconsistent; i.e., in general $F(P)$ is not a Λ-valuation and hence not a Λ-judgment.*

We are interested in the *average rule* F_{av}, specified by

$$F_{\text{av}}(P)(\phi) = \frac{\sum_{i \in N} J_i(\phi)}{|N|},$$

where $P = \langle J_i \rangle_{i \in N}$ and $\phi \in \mathcal{A}$.

Classical judgment aggregation is covered in the many-valued setting by modeling acceptance of a proposition as assigning 1 and rejection by assigning 0 to it. In other words, classical judgements and collective sets are identified with a many-valued assignment, where the truth value set $[0,1]$ is restricted to $\{0,1\}$. In particular, one can thus represent the majority rule F_{maj} as a function of type $F: \mathbf{P} \to \{0,1\}^\mathcal{A}$.

Impossibility results, like Theorem 1 of Section 2.3, have been generalized to a many-valued setting. In particular, a result of Herzberg [11][3] applies to the logics, KZ, Ł and G, that we will consider below. It considers agenda conditions that straightforwardly generalize the concepts of Definition 6 and entails that also in the corresponding many-valued setting, for non-trivial agendas, there is no non-dictatorial aggregation rule that is rational, unanimous, and systematic.

We specify a few important possible properties of aggregation rules that have been discussed in the literature; see, e.g., [9].

Definition 10. *Given a JA problem $\langle N, \mathcal{A}\rangle$ and the corresponding set $\mathbf{P}$ of judgment profiles, an aggregation rule $F : \mathbf{P} \to [0,1]^\mathcal{A}$ is*

- *unanimous iff $\forall \phi \in \mathcal{A} \ \forall P = \langle J_i \rangle_{i \in N} \in \mathbf{P}$: $[\forall i \in N \ J_i(\phi) = 1]$ implies $F(P)(\phi) = 1$;*

- *strongly unanimous iff it is unanimous and moreover $\forall \phi \in \mathcal{A} \ \forall P = \langle J_i \rangle_{i \in N} \in \mathbf{P} : [\forall i \in N : J_i(\phi) = 0]$ implies $F(P)(\phi) = 0$;*

- *anonymous iff for any permutation $\sigma : N \to N$ we have $\forall \langle J_i \rangle_{i \in N} \in \mathbf{P} : F(\langle J_i \rangle_{i \in N}) = F(\langle J_{\sigma(i)} \rangle_{i \in N})$;*

---

[3] We refer to [11] for a formal presentation of the corresponding theorem and to [7] for an interesting generalization of it.

- monotonic *iff* $\forall \phi \in \mathcal{A} \; \forall \langle J_i \rangle_{i \in N}, \langle J'_i \rangle_{i \in N} \in \mathbf{P}$: $[\forall i \in N : J_i(\phi) \leq J'_i(\phi)]$ *implies* $[F(\langle J_i \rangle_{i \in N})(\phi) \leq F(\langle J'_i \rangle_{i \in N})(\phi)]$;

- strongly monotonic *iff* $\forall \phi \in \mathcal{A} \; \forall \langle J_i \rangle_{i \in N}, \langle J'_i \rangle_{i \in N} \in \mathbf{P}$: $[\forall i \in N : J_i(\phi) < J'_i(\phi)]$ *implies* $[F(\langle J_i \rangle_{i \in N})(\phi) < F(\langle J'_i \rangle_{i \in N})(\phi)]$;

- systematic *iff* $\exists f : [0,1]^n \to [0,1]$ *s.t.* $\forall \langle J_i \rangle_{i \in N} \in \mathbf{P} \; \forall \phi \in \mathcal{A} : F(\langle J_i \rangle_{i \in N})(\phi) = f(J_1(\phi), \ldots, J_n(\phi))$

Remark 3. *Unanimousity is usually formulated only for classical (Boolean) judgment aggregation. Anonymity and systematicity are also frequently used in the literature on classical judgment aggregation (see, e.g., [9]) and have been generalized to many-valued judgment aggregation (see, e.g., [5] and [7]). The classical version of monotonicity states that if one agent switches from rejecting to accepting a collectively accepted proposition (keeping all other individual judgments), this proposition should remain collectively accepted (see, e.g., [3] for a discussion of this condition.) This property is entailed by the above definition of monotonicity for judgments to $\{0,1\}$. The many-valued versions of (strong) monotonicity in Definition 10 have been first formulated in [21].*

3.2 Many-valued aggregation for the discursive dilemma

Let us revisit the *discursive dilemma* (Section 2.1) in terms of the terminology introduced in Section 3.1. It can be represented in the context of the judgment aggregation problem $\langle N, \mathcal{A} \rangle$, where $N = \{1,2,3\}$ and $\mathcal{A} = \{p, q, p \wedge q\}$. More specifically, the dilemma arises for the judgment profile $P = \langle J_i \rangle_{i \in N}$, where

$$J_1(p) = 1, \quad J_1(q) = 1, \quad J_1(p \wedge q) = 1,$$
$$J_2(p) = 1, \quad J_2(q) = 0, \quad J_2(p \wedge q) = 0,$$
$$J_3(p) = 0, \quad J_3(q) = 1, \quad J_3(p \wedge q) = 0.$$

J_1, J_2, and J_3 are consistent (with respect to classical logic), but F_{maj} yields the following inconsistent valuation:

$$F_{\text{maj}}(P)(p) = 1, \; F_{\text{maj}}(P)(q) = 1, \; F_{\text{maj}}(P)(p \wedge q) = 0.$$

If we apply the aggregation rule F_{av} instead of F_{maj}, we obtain

$$F_{\text{av}}(P)(p) = \tfrac{2}{3}, \; F_{\text{av}}(P)(q) = \tfrac{2}{3}, \; F_{\text{av}}(P)(p \wedge q) = \tfrac{1}{3}.$$

Whether this corresponds to a consistent many-valued judgment depends on the applied logic. We may note that by interpreting $\wedge$ as *strong conjunction* according to Łukasiewicz logic Ł, where the corresponding truth function is the Łukasiewicz t-norm $x \circ_Ł y = \max(0, x + y - 1)$, we obtain a consistent average judgment, since $\frac{2}{3} \circ_Ł \frac{2}{3} = \frac{1}{3}$.

On the other hand, consider the judgment profile P' for the same JA problem which agrees everywhere with the foregoing P except that $J_2(p) = 0$. For P' we obtain

$$F_{\text{maj}}(P')(p) = 0, \quad F_{\text{maj}}(P')(q) = 1, \quad F_{\text{maj}}(P')(p \wedge q) = 0,$$
$$F_{\text{av}}(P')(p) = \tfrac{1}{3}, \quad F_{\text{av}}(P')(q) = \tfrac{2}{3}, \quad F_{\text{av}}(P')(p \wedge q) = \tfrac{1}{3}.$$

Clearly, $F_{\text{maj}}(P')$ amounts to a classical valuation. In contrast, $F_{\text{av}}(P')$ is not Ł-consistent if we interpret conjunction as the Łukasiewicz t-norm. However, another popular truth function for conjunction, namely assigning the *minimum* of the values of the conjuncts to the conjunction, yields a consistent many-valued valuation since $\min(\tfrac{1}{3}, \tfrac{2}{3}) = \tfrac{1}{3}$.

The above analysis illustrates the negative message conveyed by the general impossibility result alluded to at the end of Section 3.1: one cannot expect averaging (which, of course, is anonymous, unanimous, systematic, and in general non-dictatorial) to yield consistent collective many-valued judgments in general. However, there is also a positive message: both many-valued interpretations of conjunction considered above allow consistent aggregation of certain types of individual judgments. In the rest of the paper, we will demonstrate that these positive observations about the discursive dilemma can be generalized to subsume a much wider range of agendas and judgment profiles.

4 Consistent average aggregation

4.1 Average aggregation of classical judgment profiles using KZ

In Section 3.2 we considered a judgment profile P' for the agenda $\mathcal{A} = \{p, q, p \wedge q\}$ that yields a consistent majority judgment. More interestingly, F_{av}, which for each agenda item outputs the average of the individual judgments, consistently aggregates P' if we interpret conjunction as the minimum.

A simple and popular form of fuzzy logic features the truth functions $\min(x, y)$, $\max(x, y)$, and $1 - x$ for conjunction, disjunction, and negation, respectively (see, e.g., the widely used textbook [19]). Following [1], we will refer to this logic as *Kleene-Zadeh logic* KZ.

Definition 11 (Kleene-Zadeh logic KZ). *Like Boolean formulas, KZ formulas are built up from atoms At (i.e., propositional variables) using the connectives $\wedge$, $\vee$, and $\neg$. The semantics of KZ is given by extending any assignment (valuation)*

$v : At \to [0,1]$ *from atoms to arbitrary formulas as follows:* $v(\neg \phi) = 1 - v(\phi)$, $v(\phi \wedge \psi) = \min(v(\phi), v(\psi))$, $v(\phi \vee \psi) = \max(v(\phi), v(\psi))$.

Motivated by the above observations, we ask for which classical judgment profiles F_{av} yields KZ-consistent collective judgments and whether in these cases the majority judgment given by F_{maj} is consistent as well. In order to formulate a corresponding criterion we introduce the following notations.

Definition 12. *Let* $P = \langle J_i \rangle_{i \in N}$ *be a classical judgment profile for the JA problem* $\langle N, \mathcal{A} \rangle$ *and let* p, q *be propositional variables, i.e. atoms, that are elements of* $\mathcal{A}$. *For* $u, v \in \{0, 1\}$ *we define*

$$P^{\#}_{uv}(p, q) = |\{i \in N : J_i(p) = u, J_i(q) = v\}|.$$

Note that $P^{\#}_{uv}(p, q)$ just counts the number of individual judgments where p is assigned the truth value u and q is assigned the truth value v.

Definition 13. *Let* $P = \langle N, \mathcal{A} \rangle$ *be a classical judgment profile where* $\mathcal{A}$ *is closed.* P *is smooth iff, for every pair* (p, q) *of propositional variables, i.e., atoms* $\in \mathcal{A}$, $P^{\#}_{01}(p, q) = 0$ *or* $P^{\#}_{10}(p, q) = 0$.

Informally, smooth judgment profiles can be described as those where the agents do not have opposite views about pairs of atomic propositions. More precisely, the agents might disagree about whether individual propositions are true; but for every pair of atomic propositions all agents that judge one of the two propositions to be true and the other one to be false agree upon which of the two is the false and which is the true one. Consequently, one of the four possible truth value assignments for every pair of atoms is excluded. Note that the judgment profile for the discursive dilemma is not smooth. In contrast, the variation P' that was considered in Section 3.2 for the same agenda is smooth.

Smoothness is not sufficient to guarantee the KZ-consistency of F_{av}-aggregated judgments over arbitrary agendas. To see this, consider the following JA scenario with two judges. (The profile is trivially smooth, since there is only one propositional variable p.)

	p	$\neg p$	$p \wedge \neg p$
J_1:	0	1	0
J_2:	1	0	0
F_{av}	$\frac{1}{2}$	$\frac{1}{2}$	0

Clearly, the assignment F_{av} of truth values to the formulas of the agenda $\{p, \neg p, p \wedge \neg p\}$ does not constitute a KZ-valuation, since $\min(\frac{1}{2}, \frac{1}{2}) = \frac{1}{2}$, while both agents, and hence also F_{av}, assign 0 (false) to $p \wedge \neg p$.

As we will see below, the following condition characterizes agendas for which smooth judgment profiles can be aggregated KZ-consistently using F_{av}.

Definition 14. *An agenda $\mathcal{A}$ is* internally positive *iff each formula in $\mathcal{A}$ is either negation-free (i.e. without occurrences of $\neg$) or it is of the form $\neg\phi$, where ϕ is negation-free.*

Theorem 2. *Let $\langle N, \mathcal{A} \rangle$ be an internally positive and closed judgment aggregation problem. Then $F_{\text{av}}(P)$ is a KZ-valuation for every smooth classical judgment profile P for $\langle N, \mathcal{A} \rangle$.*

Theorem 3. *Let $\langle N, \mathcal{A} \rangle$ be an internally positive and closed judgment aggregation problem, where $|N|$ is an odd number. Then $F_{\text{maj}}(P)$ is (classically) consistent for every smooth classical judgment profile P for $\langle N, \mathcal{A} \rangle$.*

Theorems 2 and 3 will turn out to be corollaries of the more general Theorem 4 below.

4.2 Average aggregation of many-valued judgment profiles using KZ

We now turn to scenario (2) mentioned in Section 3.1. More precisely, we consider many-valued judgment profiles, where the individual judgments are KZ-valuations. Like for scenario (1), treated in Section 4.1, we aim at a criterion for judgment profiles P that guarantees the KZ-consistency of $F_{\text{av}}(P)$ for a wide range of agendas.

Definition 15 (Order compatible profiles). *Let $P = \langle J_i \rangle_{i \in N}$ be a KZ-judgment profile for the JA problem $\langle N, \mathcal{A} \rangle$, where $\mathcal{A}$ is closed. P is* order compatible *iff there exists an enumeration $\langle p_1, p_2, \ldots, p_m \rangle$ of all atoms (propositional variables) occurring in $\mathcal{A}$, such that*

$$J_i(p_1) \leq J_i(p_2) \leq \ldots \leq J_i(p_m)$$

for every $i \in N$.

Note that the profile of the discursive dilemma presented in Section 2.1 is not order compatible. However, the following variation of it is order compatible:

	v	b	$v \wedge b$
$J_1:$	1	1	1
$J_2:$	0	1	0
$J_3:$	0	0	0

Informally speaking, a judgment profile is order compatible if the agents agree on the *relative* degree of truth that they assign to the atomic propositions of the agenda. The following lemma relies on the fact that any such order of atomic formulas uniquely extends to negation-free formulas.

Lemma 1. *Let $P = \langle J_i \rangle_{i \in N}$ be an order compatible KZ-judgment profile for the JA problem $\langle N, \mathcal{A} \rangle$, where $\mathcal{A}$ is closed and negation-free. Then there exists an enumeration $\langle \phi_1, \phi_2, \ldots, \phi_m \rangle$ of all formulas occurring in $\mathcal{A}$, such that*

$$J_i(\phi_1) \leq J_i(\phi_2) \leq \ldots \leq J_i(\phi_m)$$

for every $i \in N$.

Proof. We proceed by induction on the logical complexity of $\mathcal{A}$, i.e., the total number k of occurrences of logical connectives in the formulas of $\mathcal{A}$. If $k = 0$, i.e. if all formulas in $\mathcal{A}$ are atomic, then the claim holds by definition of order compatibility. If $k > 0$ we have to distinguish the following two cases. (Since $\mathcal{A}$ is negation-free, $\wedge$ and $\vee$ are the only connectives to consider.)

- $\phi \wedge \psi \in \mathcal{A}$: By the induction hypothesis we have either $J_i(\phi) \leq J_i(\psi)$ for every $i \in N$ or $J_i(\psi) \leq J_i(\phi)$ for every $i \in N$. Since each judgment J_i is a KZ-valuation, we have $J_i(\phi \wedge \psi) = \min(J_i(\phi), J_i(\psi))$, which, by the above observation, is either $J_i(\phi)$ for every $i \in N$ or $J_i(\psi)$ for every $i \in N$. Hence, the shared order includes also the judgments for $\phi \wedge \psi$.

- $\phi \vee \psi \in \mathcal{A}$: The argument is analogous to that for $\phi \wedge \psi$: since $J_i(\phi \vee \psi) = \max(J_i(\phi), J_i(\psi))$, the induction hypothesis yields that $J_i(\phi \vee \psi) = J_i(\phi)$ for every $i \in N$ or $J_i(\phi \vee \psi) = J_i(\psi)$ for every $i \in N$. Consequently, the shared order includes $\phi \vee \psi$. □

Theorem 4. *Let $\langle N, \mathcal{A} \rangle$ be a JA problem, where the agenda $\mathcal{A}$ is closed and internally positive. Then the average rule yields a KZ-consistent collective judgment $F_{\mathrm{av}}(P)$ for every order compatible KZ-judgment profile P.*

Proof. Let $\langle N, \mathcal{A} \rangle$ be a JA problem with a closed and internally positive agenda $\mathcal{A}$, $n := |N|$, and let $P = \langle J_i \rangle_{i \in N}$ be an order compatible KZ-judgment profile for $\langle N, \mathcal{A} \rangle$.

First, we deal with the case that $\mathcal{A}$ is negation-free and show that $F_{\mathrm{av}}(P)$ is KZ-consistent. For this we define $\mathcal{A}_k$ for $k \in \mathbb{N}$ as follows:

- Let $\mathcal{A}_0$ be the set of atoms appearing in $\mathcal{A}$.

- For $k \in \mathbb{N}$, $\mathcal{A}_k$ is defined inductively as

$$\mathcal{A}_{k+1} := \mathcal{A}_k \cup \Big(\mathcal{A} \cap (\{\phi \wedge \psi \mid \phi, \psi \in \mathcal{A}_k\} \cup \{\phi \vee \psi \mid \phi, \psi \in \mathcal{A}_k\})\Big).$$

By construction, for $k \in \mathbb{N}$ we have that $\mathcal{A}_k \subseteq \mathcal{A}_{k+1}$, $\mathcal{A}_k \subseteq \mathcal{A}$, and that $\mathcal{A}_k$ is closed. Since $\mathcal{A}_0$ contains only atoms and since in the inductive definition no negation is introduced, it follows that for any $k \in \mathbb{N}$, $\mathcal{A}_k$ is negation-free. Moreover, for each $k \in \mathbb{N}$, let $P_k := \langle J_1|_{\mathcal{A}_k}, \ldots, J_n|_{\mathcal{A}_k}\rangle$ be the subprofile of P restricted to the formulas contained in $\mathcal{A}_k$. I.e., for each agent $i \in N$, P_k contains exactly the judgments of P that judge over the formulas of the restricted agenda $\mathcal{A}_k$. Since P is order compatible, P_k is an order compatible KZ-judgment profile for the KZ-judgment subproblem $\langle N, \mathcal{A}_k \rangle$. This follows from the fact that for any $k \in \mathbb{N}$, the atoms $\mathcal{A}_0 \subseteq \mathcal{A}_k$ and the individual judgments for each atom of P_k stay the same as in P.

By induction over k we show that $F_{\text{av}}(P_k)$ is KZ-consistent for any $k \in \mathbb{N}$. The base case for $k = 0$ is trivial, as all formulas in $\mathcal{A}_0$ are atomic and any assignment of values in $[0,1]$ to atoms constitutes a valuation over any logic.

We continue with the induction step from $\mathcal{A}_k$ to $\mathcal{A}_{k+1}$. Suppose that $F_{\text{av}}(P_k)$ is KZ-consistent with respect to $\langle N, \mathcal{A}_k\rangle$. We show that this implies that $F_{\text{av}}(P_{k+1})$ is KZ-consistent with respect to $\langle N, \mathcal{A}_{k+1}\rangle$.

By definition, P_k and P_{k+1} are the judgment subprofiles of P restricted to the formulas in $\mathcal{A}_k$ and $\mathcal{A}_{k+1}$, respectively, with $\mathcal{A}_k \subseteq \mathcal{A}_{k+1}$. So for every formula $\xi \in \mathcal{A}_k$, we have that for every $i \in N$, $J_i|_{\mathcal{A}_{k+1}}(\xi) = J_i|_{\mathcal{A}_k}(\xi)$ and thus

$$F_{\text{av}}(P_{k+1})(\xi) = \frac{1}{n}\sum_{i\in N} J_i|_{\mathcal{A}_{k+1}}(\xi) = \frac{1}{n}\sum_{i\in N} J_i|_{\mathcal{A}_k}(\xi) = F_{\text{av}}(P_k)(\xi).$$

Therefore no inconsistency can arise from formulas in $\mathcal{A}_k$ in the average aggregation of P_{k+1} with respect to $\langle N, \mathcal{A}_{k+1}\rangle$.

Next, we show that the unique KZ-consistent extension of $F_{\text{av}}(P_{k+1})|_{\mathcal{A}_k}$ to $\mathcal{A}_{k+1}$ coincides with $F_{\text{av}}(P_{k+1})$. Let $\xi \in \mathcal{A}_{k+1} \setminus \mathcal{A}_k$ be any formula not already contained in $\mathcal{A}_k$. We proceed with a case distinction.

$\xi = \phi \wedge \psi$: By construction of $\mathcal{A}_{k+1}$, it follows that $\phi, \psi \in \mathcal{A}_k$. Since P_k is an order compatible KZ-judgment profile for the restricted subproblem $\langle N, \mathcal{A}_k\rangle$, where $\mathcal{A}_k$ is negation-free, it follows from Lemma 1 that there is an enumeration $\langle \phi_1, \phi_2, \ldots, \phi_m\rangle$ of all formulas occurring in $\mathcal{A}_k$, such that for every $i \in N$:

$$J_i|_{\mathcal{A}_k}(\phi_1) \leq J_i|_{\mathcal{A}_k}(\phi_2) \leq \cdots \leq J_i|_{\mathcal{A}_k}(\phi_m)$$

Since $\phi, \psi \in \mathcal{A}_k$, either for all $i \in N$, $J_i|_{\mathcal{A}_k}(\phi) \leq J_i|_{\mathcal{A}_k}(\psi)$ or for all $i \in N$, $J_i|_{\mathcal{A}_k}(\psi) \leq J_i|_{\mathcal{A}_k}(\phi)$. Let us assume that for all $i \in N$, $J_i|_{\mathcal{A}_k}(\phi) \leq J_i|_{\mathcal{A}_k}(\psi)$, the other case can be handled analogously. Since $\phi, \psi \in \mathcal{A}_k$, it holds that $J_i|_{\mathcal{A}_{k+1}}(\phi) = J_i|_{\mathcal{A}_k}(\phi) \leq J_i|_{\mathcal{A}_k}(\psi) = J_i|_{\mathcal{A}_{k+1}}(\psi)$ for every $i \in N$. Thus we obtain

$$F_{\mathrm{av}}(P_{k+1})(\phi) = \frac{1}{n} \sum_{i \in N} J_i|_{\mathcal{A}_{k+1}}(\phi) \leq \frac{1}{n} \sum_{i \in N} J_i|_{\mathcal{A}_{k+1}}(\psi) = F_{\mathrm{av}}(P_{k+1})(\psi).$$

From this it follows that the value for the consistent extension is

$$\min(F_{\mathrm{av}}(P_{k+1})(\phi), F_{\mathrm{av}}(P_{k+1})(\psi)) = F_{\mathrm{av}}(P_{k+1})(\phi).$$

Moreover, for each $i \in N$, J_i and therefore also its restriction $J_i|_{\mathcal{A}_{k+1}}$ to $\mathcal{A}_{k+1}$ is a KZ-consistent judgment. Therefore, $J_i|_{\mathcal{A}_{k+1}}(\phi \wedge \psi) = \min(J_i|_{\mathcal{A}_{k+1}}(\phi), J_i|_{\mathcal{A}_{k+1}}(\psi)) = J_i|_{\mathcal{A}_{k+1}}(\phi)$ for all $i \in N$. Thus it follows that

$$F_{\mathrm{av}}(P_{k+1})(\phi \wedge \psi) = \frac{1}{n} \sum_{i \in N} J_i|_{\mathcal{A}_{k+1}}(\phi \wedge \psi) = \frac{1}{n} \sum_{i \in N} J_i|_{\mathcal{A}_{k+1}}(\phi) = F_{\mathrm{av}}(P_{k+1})(\phi),$$

which coincides with the value of the extension as desired.

$\xi = \phi \vee \psi$: The argument is analogous to that for $\xi = \phi \wedge \psi$.

This concludes the inductive part of the proof. We further have to show that there is some $k \in \mathbb{N}$ such that for all $l \geq k$, $\mathcal{A}_l = \mathcal{A}$, i.e., that $F_{\mathrm{av}}(P)$ is KZ-consistent with respect to $\langle N, \mathcal{A} \rangle$.

By definition $\mathcal{A}$ is finite and thus the chain $\mathcal{A}_0 \subseteq \mathcal{A}_1 \subseteq \mathcal{A}_2 \subseteq \cdots \subseteq \mathcal{A}$ eventually converges to $\mathcal{A}$. I.e., there is some $k \in \mathbb{N}$ such that for all $l \in \mathbb{N}$ where $l > k$, $\mathcal{A}_l = \mathcal{A}_k \subseteq \mathcal{A}$. In order to obtain a contradiction, we assume that for this $k \in \mathbb{N}$, $\mathcal{A}_k \subsetneq \mathcal{A}$. So there must be a formula $\xi \in \mathcal{A} \setminus \mathcal{A}_k$ of minimal logical complexity, i.e., minimal number of occurrences of logical connectives of any formula in $\mathcal{A} \setminus \mathcal{A}_k$. Since $\mathcal{A}_0 \subseteq \mathcal{A}_k$, it follows that ξ cannot be an atom. Thus $\xi = \phi \circ \psi$ with $\circ \in \{\wedge, \vee\}$. Since $\mathcal{A}$ is closed, we have that $\phi, \psi \in \mathcal{A}$. Since ϕ and ψ are less complex than ξ, they cannot belong to $\mathcal{A} \setminus \mathcal{A}_k$ (since ξ is a minimally complex formula of this kind). Therefore $\phi, \psi \in \mathcal{A}_k$. However, $\xi \in \mathcal{A}$ is of minimal logical complexity with respect to the formulas in $\mathcal{A} \setminus \mathcal{A}_k$, so $\xi = \phi \circ \psi \in \mathcal{A}_{k+1} = \mathcal{A}_k$ by the inductive construction of $\mathcal{A}_k$ and the fact that the chain stabilizes. However, this contradicts the assumption that $\xi \notin \mathcal{A}_k$ and hence we obtain $\mathcal{A}_k = \mathcal{A}$. It follows that $P_k = P$ and therefore $F_{\mathrm{av}}(P)$ is KZ-consistent.

This concludes the proof for the case where $\mathcal{A}$ is negation-free. The final step of the proof is now to generalize the above result to internally positive agendas.

Let $\langle N, \mathcal{A} \rangle$ be a KZ-judgment aggregation problem where $\mathcal{A}$ is closed and internally positive; $n := |N|$. We set $\mathcal{A}'$ to contain exactly the negation-free formulas of $\mathcal{A}$, in particular, $\mathcal{A}'$ has the same set of atoms as $\mathcal{A}$. Since $\mathcal{A}$ is internally positive, by Definition 14 it can be written as $\mathcal{A} = \mathcal{A}' \cup \{\neg \phi \mid \phi \in \mathcal{B}\}$ where $\mathcal{B}$ contains only negation-free formulas. Since $\mathcal{A}$ is closed, we have that $\mathcal{B} \subseteq \mathcal{A}$. $\mathcal{A}'$ contains all negation-free formulas of $\mathcal{A}$ and thus $\mathcal{B} \subseteq \mathcal{A}'$. Moreover, due to the closedness of $\mathcal{A}$ it follows that $\mathcal{A}'$ is also closed. Let $P = \langle J_1, \ldots, J_n \rangle$ be an order compatible KZ-judgment profile for $\langle N, \mathcal{A} \rangle$ and let $P' = \langle J_1|_{\mathcal{A}'}, \ldots, J_n|_{\mathcal{A}'} \rangle$ be its restriction to $\mathcal{A}'$. Since P is order compatible and P' restricts the judgments to a subset of formulas, P' is also an order compatible KZ-judgment profile for $\langle N, \mathcal{A}' \rangle$ with the closed negation-free agenda $\mathcal{A}'$. Thus, by what was shown above, $F_{\text{av}}(P')$ is KZ-consistent. Since $\mathcal{A}' \subseteq \mathcal{A}$, we have for every formula $\xi \in \mathcal{A}'$

$$F_{\text{av}}(P)|_{\mathcal{A}'}(\xi) = \frac{1}{n} \sum_{i \in N} J_i|_{\mathcal{A}'}(\xi) = F_{\text{av}}(P')(\xi).$$

Therefore no inconsistency can arise from formulas in $\mathcal{A}'$ in the average aggregation of P with respect to $\langle N, \mathcal{A} \rangle$. It remains to show how this can be extended to formulas $\neg \phi$ for $\phi \in \mathcal{B}$. For every $i \in N$, J_i is consistent and thus $J_i(\neg \phi) = 1 - J_i(\phi)$. Thus it follows that

$$F_{\text{av}}(P)(\neg \phi) = \frac{1}{n} \sum_{i \in N} (1 - J_i(\phi)) = 1 - \frac{1}{n} \sum_{i \in N} J_i(\phi) = 1 - F_{\text{av}}(P)(\phi),$$

which is still KZ-consistent. This concludes the proof. $\square$

To verify that Theorem 2 is entailed by Theorem 4, as claimed in Section 4.1, it suffices to observe that order compatibility is entailed by smoothness for classical (Boolean) judgment profiles. If the judgment profile P consists of classical judgments J_i for $i \in N$ (i.e., if $J_i(\phi) \in \{0, 1\}$ for every $\phi \in \mathcal{A}$), and if $\langle p_1, \ldots, p_m \rangle$ is an enumeration of (some of the) atoms of $\mathcal{A}$ such that $J_i(p_1) \leq \cdots \leq J_i(p_m)$ for all $i \in N$, and $q \in \mathcal{A}$ is another atom, then there are two cases. Either there is $j \in \{1, \ldots, m\}$ such that $J_i(p_j) = J_i(q)$ for all $i \in N$; then we extend the enumeration by inserting the atom q directly after p_j. Otherwise, by smoothness of P, for each $j \in \{1, \ldots, m\}$ we have $P^{\#}_{01}(p_j, q) > 0 = P^{\#}_{10}(p_j, q)$ or $P^{\#}_{01}(p_j, q) = 0 < P^{\#}_{10}(p_j, q)$. In the first case q needs to appear in the enumeration after p_j, in the latter case before p_j. Let $J = \{j \in \{1, \ldots, m\} \mid P^{\#}_{01}(p_j, q) > 0\}$. If $J = \emptyset$, place q before p_1 in the enumeration of atoms, else let $k := \max J$ and place q directly after p_k. In this way order compatibility of P can be established from smoothness by extending an initial enumeration of a single atom in finitely many steps. This means that Theorem 2 amounts to a special case of Theorem 4. It remains to show that Theorem 3 also follows from Theorem 4.

Proof of Theorem 3. Applying Theorem 4 to a JA problem with internally positive and closed agenda $\mathcal{A}$, we obtain a KZ-valuation $F_{\text{av}}(P)$ for every smooth (and hence order compatible) classical judgment profile $P = \langle N, \mathcal{A} \rangle$. A corresponding collective (Boolean) majority judgment $F_{\text{maj}}(P)$ is obtained by setting

$$F_{\text{maj}}(P)(\phi) = \begin{cases} 1 & F_{\text{av}}(P)(\phi) > 0.5, \\ 0 & \text{otherwise.} \end{cases}$$

for every $\phi \in \mathcal{A}$. Since $n = |N|$ is odd, by assumption, $F_{\text{av}}(P)(\phi)$, which is the fraction of agents that judge ϕ to be true, can never be 0.5. This fact entails that the KZ-consistency of $F_{\text{av}}(P)$ implies the classical consistency of $F_{\text{maj}}(P)$. To see this we distinguish three cases according to the logical form of the agenda items and check that $F_{\text{maj}}(P)$ indeed respects the classical truth functions.

(1) For $\neg \phi \in \mathcal{A}$ first suppose that $F_{\text{av}}(P)(\phi) > 0.5$ and therefore $F_{\text{maj}}(P)(\phi) = 1$. Since $F_{\text{av}}(P)$ is a KZ-valuation, we obtain $F_{\text{av}}(P)(\neg\phi) = 1 - F_{\text{av}}(P)(\phi) < 0.5$ and hence $F_{\text{maj}}(P)(\neg\phi) = 0$, by definition. Similarly, if $F_{\text{av}}(P)(\phi) < 0.5$ and hence $F_{\text{maj}}(P)(\phi) = 0$, then $F_{\text{av}}(P)(\neg\phi) > 0.5$ and hence $F_{\text{maj}}(P)(\neg\phi) = 1$. Summing up, we have shown that $F_{\text{maj}}(P)(\neg\phi) = 1 - F_{\text{maj}}(P)(\phi)$, as required.

(2) For $\phi \wedge \psi \in \mathcal{A}$ first suppose that $F_{\text{av}}(P)(\phi) > 0.5$ and $F_{\text{av}}(P)(\psi) > 0.5$ and therefore $F_{\text{maj}}(P)(\phi) = 1$ and $F_{\text{maj}}(P)(\psi) = 1$. Since $F_{\text{av}}(P)$ is a KZ-valuation, we obtain $F_{\text{av}}(P)(\phi \wedge \psi) = \min(F_{\text{av}}(P)(\phi), F_{\text{av}}(P)(\psi)) > 0.5$ and hence $F_{\text{maj}}(P)(\phi \wedge \psi) = 1$, by definition. Now suppose that either $F_{\text{av}}(P)(\phi) < 0.5$ or $F_{\text{av}}(P)(\psi) < 0.5$, or both. Since $F_{\text{av}}(P)$ is a KZ-valuation, this implies $F_{\text{av}}(P)(\phi \wedge \psi) < 0.5$ and hence $F_{\text{maj}}(P)(\phi \wedge \psi) = 0$, by definition. Summing up, we have shown that $F_{\text{maj}}(P)(\phi \wedge \psi) = 1$ iff $F_{\text{maj}}(P)(\phi) = 1$ and $F_{\text{maj}}(P)(\psi) = 1$, as required.

(3) The case for $\phi \vee \psi \in \mathcal{A}$ is analogous to case (2). □

4.3 Average aggregation of classical judgment profiles using Ł

In Section 3.2 we have observed that for the discursive dilemma, where majority aggregation is (classically) inconsistent, the average rule F_{av} yields an assignment of truth values in $[0, 1]$ that is consistent if conjunction is modeled by the Łukasiewicz t-norm. In this section we generalize this situation to a wider range of JA problems and Ł-judgment profiles. More precisely, we stipulate that conjunction, disjunction, and negation are interpreted using the truth functions $\max(0, x+y-1)$, $\min(1, x+y)$, and $1 - x$, respectively.[4]

Recall that the condition of smoothness, introduced in Section 4.1 to single out classical judgment profiles that guarantee KZ-consistent average aggregations for internally positive agendas, has been defined by excluding certain truth value assignments from the profile of individual judgments. The following condition is defined in a similar manner, once more employing Definition 12.

Definition 16. *A classical judgment profile P for the JA problem $\langle N, \mathcal{A}\rangle$ is called lopsided iff, for every pair p, q of distinct propositional variables contained in $\mathcal{A}$, $P_{00}^{\#}(p,q) = 0$ or $P_{11}^{\#}(p,q) = 0$.*

There is no analogue of Lemma 1 for lopsided Ł-judgment profiles: in contrast to order compatibility and hence also to smoothness, lopsidedness cannot be generalized from atomic to arbitrary positive non-atomic formulas using Ł. For a condition that guarantees the Ł-consistency of F_{av}-generated collective judgments, we introduce the following, more restrictive type of agendas.

Definition 17. *An agenda $\mathcal{A}$ is called flat iff all formulas in $\mathcal{A}$ are of the form p, $\neg p$, $p \wedge q$, $\neg(p \wedge q)$, $p \vee q$, or $\neg(p \vee q)$, where p and q are distinct propositional variables.*

Theorem 5. *Let $\langle N, \mathcal{A}\rangle$ be a JA problem, where the agenda $\mathcal{A}$ is closed and flat. Then the average rule yields a Ł-consistent collective judgment $F_{\mathrm{av}}(P)$ for every lopsided judgment profile P.*

Proof. If the agenda $\mathcal{A}$ consists only of atomic formulas, then the claim holds trivially. Since the agenda is flat, we have to consider the following cases, where p and q are distinct propositional variables occurring in compound formulas of the agenda.

$p \wedge q$: To prove that $F_{\mathrm{av}}(P)$ yields an Ł-valuation we have to show that the following condition holds:

$$F_{\mathrm{av}}(P)(p \wedge q) = \max(0, F_{\mathrm{av}}(P)(p) + F_{\mathrm{av}}(P)(q) - 1),$$

---

[4]In Ł one can express not only these truth functions, but also those of KZ. One speaks of $\max(0, x + y - 1)$ and $\min(1, x + y)$ as the truth functions for strong conjunction and strong disjunction, while $\min(x, y)$ and $\max(x, y)$ define weak conjunction and weak disjunction, respectively. (Also the terms t-norm versus lattice based conjunction and disjunction are used in this context.) Usually the symbols $\wedge$ and $\vee$ refer to weak conjunction and disjunction in Ł, cf. [10]. However, here we do not want to change the propositional language (syntax) of a given agenda, but only its semantics. Hence, we retain the symbols $\wedge$ and $\vee$, but interpret them by the Lukasiewicz t-norm and co-t-norm, respectively.

where $F_{\text{av}}(P)(\phi) = \sum_{i \in N} \frac{J_i(\phi)}{n}$ for $\phi \in \{p, q, p \wedge q\}$, $P = \langle J_i \rangle_{i \in N}$ and $n = |N|$. For a more compact presentation we abbreviate $P_{xy}^{\#}(p, q)$ by i_{xy} ($x, y \in \{0, 1\}$). By definition, $F_{\text{av}}(P)(p) = \frac{i_{10} + i_{11}}{n}$ and $F_{\text{av}}(P)(q) = \frac{i_{01} + i_{11}}{n}$. Since p and q are propositional variables, the lopsidedness of P means that we have $i_{00} = 0$ or $i_{11} = 0$. Since $n = i_{00} + i_{01} + i_{10} + i_{11}$, it is straightforward to check that in both cases we obtain

$$F_{\text{av}}(P)(p \wedge q) = \frac{i_{11}}{n} = \max(0, \frac{i_{01} + i_{11}}{n} + \frac{i_{10} + i_{11}}{n} - 1).$$

Hence the condition is satisfied.

$p \vee q$: The argument is analogous to that for $p \wedge q$.

To finish the proof it suffices to note that the case for negated formulas in $\mathcal{A}$ is exactly as in the proof of Theorem 4. □

Remark 4. *It is possible to slightly relax the flatness condition. Moreover, one can generalize lopsidedness to cover certain many-valued judgment profiles, thus leading to a generalization of Theorem 5 that applies to scenario (2) described in Section 3.1. However, the corresponding formulations are rather technical without providing much additional insight. Hence they are omitted here.*

5 Possibility for logic KZ

In Section 4 we established a possibility theorem for Kleene-Zadeh logic KZ using average aggregation. In the following section we generalize this result to a wider range of aggregation functions. For this purpose, consider the following definition.

Definition 18. *Let f be a function of type $[0, 1]^n \to [0, 1]$ for some $n \geq 1$. We call f monotonic iff the following holds: $[\forall i \in \{1, \ldots, n\}\ x_i \leq x_i']$ implies $f(x_1, \ldots, x_n) \leq f(x_1', \ldots, x_n')$. Similarly, f is called strongly monotonic iff the following holds: $[\forall i \in \{1, \ldots, n\}\ x_i < x_i']$ implies $f(x_1, \ldots, x_n) < f(x_1', \ldots, x_n')$.*

Lemma 2. *Let F be a (strongly) monotonic and systematic aggregation function for a JA problem $\langle N, \mathcal{A} \rangle$ with an atom $q \in \mathcal{A}$. Then any function $f : [0, 1]^{|N|} \to [0, 1]$ witnessing systematicity of F as per $\forall \langle J_i \rangle_{i \in N} \in \mathbf{P}\ \forall \phi \in \mathcal{A} : F(\langle J_i \rangle_{i \in N})(\phi) = f(J_1(\phi), \ldots, J_{|N|}(\phi))$ is strongly monotonic.*

Proof. Let F be a monotonic and systematic aggregation function and let $\langle N, \mathcal{A} \rangle$ be an arbitrary judgment aggregation problem with $n := |N|$. Since F is systematic, there is a function such that for every $\phi \in \mathcal{A}$ and every $\langle J_i \rangle_{i \in N} \in \mathbf{P}$, $F(\langle J_i \rangle_{i \in N})(\phi) = f(J_1(\phi), \ldots, J_n(\phi))$.

In order to show that any such f is monotonic, let $x_i, x'_i \in [0,1]$ be such that $x_i \leq x'_i$ holds for all $i \in N$. Define profiles $P = \langle J_i \rangle_{i \in N}$ and $P' = \langle J'_i \rangle_{i \in N}$ of $\langle N, \mathcal{A} \rangle$ by setting $J_i(p) := x_i$ and $J'_i(p) := x'_i$ for all $i \in N$ and for each atom p appearing in some formula of $\mathcal{A}$; then extend this to consistent judgments J_i and J'_i on $\mathcal{A}$. Since $J_i(p) = x_i \leq x'_i = J'_i(p)$ holds for all $i \in N$ and $p \in \mathcal{A}$, the monotonicity and systematicity of F entail

$$f(x_1, \ldots, x_n) = f(J_1(p), \ldots, J_n(p)) = F(P)(p)$$
$$\leq F(P')(p) = f(J'_1(p), \ldots, J'_n(p)) = f(x'_1, \ldots, x'_n).$$

Hence f is monotonic.

The case for strong monotonicity is analogous. $\square$

Theorem 6. *Let $\langle N, \mathcal{A} \rangle$ be a JA problem, where $\mathcal{A}$ is closed and negation-free. Then, every systematic and monotonic aggregation function F yields a KZ-consistent collective judgment $F(P)$ for every order compatible KZ-judgment profile P.*

Proof. Let $\langle N, \mathcal{A} \rangle$ be a JA problem with a closed and negation-free agenda $\mathcal{A}$, $n := |N|$, and let $P = \langle J_i \rangle_{i \in N}$ be an order compatible KZ-judgment profile for $\langle N, \mathcal{A} \rangle$. Moreover, let F be a systematic and monotonic aggregation function.

We show that $F(P)$ is KZ-consistent. For this we define $\mathcal{A}_k$ for $k \in \mathbb{N}$ as follows:

- Let $\mathcal{A}_0$ be the set of atoms appearing in $\mathcal{A}$.

- For $k \in \mathbb{N}$, $\mathcal{A}_k$ is defined inductively as

$$\mathcal{A}_{k+1} := \mathcal{A}_k \cup \Big(\mathcal{A} \cap (\{\phi \wedge \psi \mid \phi, \psi \in \mathcal{A}_k\} \cup \{\phi \vee \psi \mid \phi, \psi \in \mathcal{A}_k\}) \Big).$$

By construction, for $k \in \mathbb{N}$ we have that $\mathcal{A}_k \subseteq \mathcal{A}_{k+1}$, $\mathcal{A}_k \subseteq \mathcal{A}$, and that $\mathcal{A}_k$ is closed. Since $\mathcal{A}_0$ contains only atoms and since in the inductive definition no negation is introduced, it follows that for any $k \in \mathbb{N}$, $\mathcal{A}_k$ is negation-free. Moreover, for each $k \in \mathbb{N}$, let $P_k := \langle J_1|_{\mathcal{A}_k}, \ldots, J_n|_{\mathcal{A}_k} \rangle$ be the subprofile of P restricted to the formulas contained in $\mathcal{A}_k$. I.e., for each agent $i \in N$, P_k contains exactly the judgments of P that judge over the formulas of the restricted agenda $\mathcal{A}_k$. That P is order compatible entails that P_k is also an order compatible KZ-judgment profile for the KZ-judgment subproblem $\langle N, \mathcal{A}_k \rangle$. This follows from the fact that for any $k \in \mathbb{N}$, the atoms $\mathcal{A}_0 \subseteq \mathcal{A}_k$ and the individual judgments for each atom of P_k stay the same as in P

By systematicity, we have $F(P)(\phi) = f(J_1(\phi), \ldots, J_n(\phi))$ for all $\phi \in \mathcal{A}$ and some $f : [0,1]^n \to [0,1]$, which is monotonic by Lemma 2 and closedness of $\mathcal{A}$. This allows us to define a judgment aggregation function F_k on profiles $P' = \langle J'_1, \ldots, J'_n \rangle$ of $\langle N, \mathcal{A}_k \rangle$ by setting $F_k(P')(\phi) := f(J'_1(\phi), \ldots, J'_n(\phi))$ for each $\phi \in \mathcal{A}_k$. In particular, we have $F_k(P_k)(\phi) = f(J_1|_{\mathcal{A}_k}(\phi), \ldots, J_n|_{\mathcal{A}_k}(\phi)) = f(J_1(\phi), \ldots, J_n(\phi)) = F(P)(\phi)$ for $\phi \in \mathcal{A}_k$, i.e., $F_k(P_k) = F(P)|_{\mathcal{A}_k}$.

By induction over k we show that $F_k(P_k)$ is KZ-consistent for any $k \in \mathbb{N}$. The base case for $k = 0$ is trivial, as all formulas in $\mathcal{A}_0$ are atomic and any assignment of values in $[0,1]$ to atoms constitutes a valuation over any logic.

We continue with the induction step from $\mathcal{A}_k$ to $\mathcal{A}_{k+1}$. Let $k \in \mathbb{N}$ and suppose that $F_k(P_k)$ is KZ-consistent with respect to $\langle N, \mathcal{A}_k \rangle$. We show that then $F_{k+1}(P_{k+1})$ is KZ-consistent with respect to $\langle N, \mathcal{A}_{k+1} \rangle$.

By definition, P_k and P_{k+1} are the judgment subprofiles of P restricted to the formulas in $\mathcal{A}_k$ and $\mathcal{A}_{k+1}$, respectively, with $\mathcal{A}_k \subseteq \mathcal{A}_{k+1}$. So for every formula $\xi \in \mathcal{A}_k$, we have that $F_{k+1}(P_{k+1})(\xi) = F(P)(\xi) = F_k(P_k)(\xi)$, i.e., $F_{k+1}(P_{k+1})|_{\mathcal{A}_k} = F_k(P_k)$. Therefore no inconsistency can arise from formulas in $\mathcal{A}_k$ in the aggregation of P_{k+1} with respect to $\langle N, \mathcal{A}_{k+1} \rangle$.

Next, we show that the unique KZ-consistent extension of $F_{k+1}(P_{k+1})|_{\mathcal{A}_k}$ to $\mathcal{A}_{k+1}$ coincides with $F_{k+1}(P_{k+1})$. Let $\xi \in \mathcal{A}_{k+1} \setminus \mathcal{A}_k$ be any formula not already contained in $\mathcal{A}_k$. We proceed with a case distinction.

$\xi = \phi \wedge \psi$: By construction of $\mathcal{A}_{k+1}$, it follows that $\phi, \psi \in \mathcal{A}_k$. Since P_k is an order compatible KZ-judgment profile for the restricted subproblem $\langle N, \mathcal{A}_k \rangle$, where $\mathcal{A}_k$ is closed and negation-free, it follows from Lemma 1 that there is an enumeration $\langle \phi_1, \phi_2, \ldots, \phi_m \rangle$ of all formulas occurring in $\mathcal{A}_k$, such that for every $i \in N$:
$$J_i|_{\mathcal{A}_k}(\phi_1) \leq J_i|_{\mathcal{A}_k}(\phi_2) \leq \cdots \leq J_i|_{\mathcal{A}_k}(\phi_m)$$

Since $\phi, \psi \in \mathcal{A}_k$, either for all $i \in N$, $J_i|_{\mathcal{A}_k}(\phi) \leq J_i|_{\mathcal{A}_k}(\psi)$ or for all $i \in N$, $J_i|_{\mathcal{A}_k}(\psi) \leq J_i|_{\mathcal{A}_k}(\phi)$. Let us assume that for all $i \in N$, $J_i|_{\mathcal{A}_k}(\phi) \leq J_i|_{\mathcal{A}_k}(\psi)$. The other case can be handled analogously. Since $\phi, \psi \in \mathcal{A}_k$, it holds that $J_i|_{\mathcal{A}_{k+1}}(\phi) = J_i|_{\mathcal{A}_k}(\phi) \leq J_i|_{\mathcal{A}_k}(\psi) = J_i|_{\mathcal{A}_{k+1}}(\psi)$ for every $i \in N$. Since f is a monotonic function and $J_i|_{\mathcal{A}_{k+1}}(\phi) \leq J_i|_{\mathcal{A}_{k+1}}(\psi)$, we obtain

$$F_{k+1}(P_{k+1})(\phi) = f(J_1|_{\mathcal{A}_{k+1}}(\phi), \ldots, J_n|_{\mathcal{A}_{k+1}}(\phi))$$
$$\leq f(J_1|_{\mathcal{A}_{k+1}}(\psi), \ldots, J_n|_{\mathcal{A}_{k+1}}(\psi)) = F_{k+1}(P_{k+1})(\psi).$$

Hence it follows that the value for the consistent extension is

$$\min(F_{k+1}(P_{k+1})(\phi), F_{k+1}(P_{k+1})(\psi)) = F_{k+1}(P_{k+1})(\phi).$$

Moreover, for each $i \in N$, J_i and therefore also the restriction $J_i|_{\mathcal{A}_{k+1}}$ to $\mathcal{A}_{k+1}$ is a KZ-consistent judgment. Therefore, $J_i|_{\mathcal{A}_{k+1}}(\phi \wedge \psi) = \min(J_i|_{\mathcal{A}_{k+1}}(\phi), J_i|_{\mathcal{A}_{k+1}}(\psi)) = J_i|_{\mathcal{A}_{k+1}}(\phi)$ for all $i \in N$. Thus it follows that

$$F_{k+1}(P_{k+1})(\phi \wedge \psi) = f(J_1|_{\mathcal{A}_{k+1}}(\phi \wedge \psi), \ldots, J_n|_{\mathcal{A}_{k+1}}(\phi \wedge \psi))$$
$$= f(J_1|_{\mathcal{A}_{k+1}}(\phi), \ldots, J_n|_{\mathcal{A}_{k+1}}(\phi)) = F_{k+1}(P_{k+1})(\phi)$$

which coincides with the value of the extension as desired.

$\xi = \phi \vee \psi$: The argument is analogous to that for $\xi = \phi \wedge \psi$.

This concludes the inductive part of the proof. We still have to show that there is some $k \in \mathbb{N}$ such that for all $l \geq k$, $\mathcal{A}_l = \mathcal{A}$, i.e., that $P_l = P$ and $F(P)$ is KZ-consistent with respect to $\langle N, \mathcal{A} \rangle$. This can be done with the same argument as given in the proof of Theorem 4.

This concludes the proof for the case that $\mathcal{A}$ is negation-free. □

Theorem 6 only applies to negation-free formulas. In order to fully generalize the result of Section 4.2, we introduce the following negation-oriented property of aggregation functions.

Definition 19 (Self-dual systematicity). *An aggregation function F for a JA-problem $\langle N, \mathcal{A} \rangle$ with $N = \{1, \ldots, n\}$ is self-dually systematic iff there is some function f with $f(1 - x_1, \ldots, 1 - x_n) = 1 - f(x_1, \ldots, x_n)$ for any $\langle x_1, \ldots, x_n \rangle$ such that $\forall P = \langle J_1, \ldots, J_n \rangle \in \mathbf{P} \; \forall \phi \in \mathcal{A} : F(P)(\phi) = f(J_1(\phi), \ldots, J_n(\phi))$.*

Remark 5. *Note that every self-dually systematic aggregation function is also systematic, as the decision function is only restricted to a subset of systematic functions.*

Corollary 1. *Let $\langle N, \mathcal{A} \rangle$ be a JA problem, where $\mathcal{A}$ is closed and internally positive. Then every self-dually systematic and monotonic aggregation function F yields a KZ-consistent collective judgment $F(P)$ for every order compatible KZ-judgment profile P for $\langle N, \mathcal{A} \rangle$.*

Proof. The inductive part for closed and negation-free agendas $\mathcal{A}$ has already been shown in Theorem 6. The final step of the proof is now to generalize the above result to internally positive agendas.

Let $\langle N, \mathcal{A} \rangle$ be a KZ-judgment aggregation problem where $\mathcal{A}$ is closed and internally positive. Let F be a self-dually systematic and monotonic aggregation function for $\langle N, \mathcal{A} \rangle$. We set $\mathcal{A}'$ to contain exactly the negation-free formulas of $\mathcal{A}$. Since $\mathcal{A}$ is internally positive, by Definition 14 it can be written as $\mathcal{A} = \mathcal{A}' \cup \{\neg \phi \mid \phi \in \mathcal{B}\}$ where $\mathcal{B}$ contains only negation-free formulas. Since $\mathcal{A}$ is closed, we have that

$\mathcal{B} \subseteq \mathcal{A}$. $\mathcal{A}'$ contains all negation-free formulas of $\mathcal{A}$ and thus $\mathcal{B} \subseteq \mathcal{A}'$. Moreover, due to the closedness of $\mathcal{A}$ it follows that $\mathcal{A}'$ is also closed. Let $P = \langle J_1, \ldots, J_n \rangle$ be an order compatible KZ-judgment profile for $\langle N, \mathcal{A} \rangle$ and let $P' = \langle J_1|_{\mathcal{A}'}, \ldots, J_n|_{\mathcal{A}'} \rangle$ be its restriction to $\mathcal{A}'$. Since P is order compatible, and P' restricts its judgments only to a subset of formulas, P' is also an order compatible KZ-judgment profile for $\langle N, \mathcal{A}' \rangle$ with the closed and negation-free agenda $\mathcal{A}'$. Let f be a self-dual and monotonic (see Lemma 2) function describing systematic aggregation by F. For any profile $P' = \langle J'_1, \ldots, J'_n \rangle$ of $\langle N, \mathcal{A}' \rangle$, we define $F'(P')$ by setting $F'(P')(\phi) := f(J'_1(\phi), \ldots, J'_n(\phi))$ for any $\phi \in \mathcal{A}'$. Now F' is systematic and monotonic, so Theorem 6 tells that $F'(P')$ is KZ-consistent. Since $\mathcal{A}' \subseteq \mathcal{A}$, we have for every formula $\xi \in \mathcal{A}'$

$$F(P)|_{\mathcal{A}'}(\xi) = f(J_1|_{\mathcal{A}'}(\xi), \ldots, J_n|_{\mathcal{A}'}(\xi)) = F'(P')(\xi).$$

Therefore no inconsistency can arise from formulas in $\mathcal{A}'$ in the aggregation of P with respect to $\langle N, \mathcal{A} \rangle$. Now we show how this can be extended to formulas $\neg \phi$ for $\phi \in \mathcal{B}$. For every $i \in N$, J_i is consistent and thus $J_i(\neg \phi) = 1 - J_i(\phi)$. Thus it follows by Definition 19 of self-dual systematicity

$$F(P)(\neg \phi) = f(J_1(\neg \phi), \ldots, J_n(\neg \phi)) = f(1 - J_1(\phi), \ldots, 1 - J_n(\phi)) = 1 - F(P)(\phi)$$

which is still KZ-consistent. This concludes the proof. $\square$

Note that, at a first glance, one might suspect that the average rule F_{av} is the only aggregation function that satisfies the condition of self-dual systematicity. However, observe that the constant aggregation function $F_{0.5}$, defined by $F_{0.5}(P)(\phi) = 0.5$ for every $P \in \mathbf{P}$ and every formula $\phi \in \mathcal{A}$, satisfies both monotonicity and self-dual systematicity. Another example, satisfying self-dual systematicity and anonymity, is the *majority rule* F_{maj}.

Example 1 (Self-dual systematicity of the majority rule). *Let $\langle N, \mathcal{A} \rangle$ be a judgment aggregation problem with a closed agenda $\mathcal{A}$ and odd $|N|$. Recall from the proof of Theorem 3 that the majority rule can be defined with reference to the average of Boolean judgments $P = \langle J_i \rangle_{i \in N}$ for $\phi \in \mathcal{A}$ as follows:*

$$F_{\mathrm{maj}}(P)(\phi) = \begin{cases} 1 & \text{if } \frac{\sum_{i \in N} J_i(\phi)}{|N|} > 0.5 \\ 0 & \text{otherwise.} \end{cases}$$

Since for odd $|N|$ and Boolean profiles it is impossible that $\frac{1}{|N|} \sum_{i \in N} J_i(\phi) = \frac{1}{2}$, one may compute the majority aggregation for such profiles also systematically as

$F_{\mathrm{maj}}(P)(\phi) = f(\langle J_i(\phi)\rangle_{i\in N})$ where

$$f(x_1,\ldots,x_n) = \begin{cases} 1 & \text{if } \frac{1}{n}\sum_{i=1}^n x_i > \frac{1}{2}, \\ \frac{1}{2} & \text{if } \frac{1}{n}\sum_{i=1}^n x_i = \frac{1}{2}, \\ 0 & \text{if } \frac{1}{n}\sum_{i=1}^n x_i < \frac{1}{2}. \end{cases}$$

In fact, this definition makes sense even if the individual judgments are not Boolean, but many-valued. In any case, we have $F_{\mathrm{maj}}(P)(\phi) = 1$ if $F_{\mathrm{av}}(P)(\phi) > 0.5$ and $F_{\mathrm{maj}}(P)(\phi) = 0$ otherwise. Let $x_1,\ldots,x_n$ be arbitrary truth values in $[0,1]$. We have observed before that $\frac{1}{n}\sum_{i=1}^n (1-x_i) = 1 - \frac{1}{n}\sum_{i=1}^n x_i$. Therefore, $\frac{1}{n}\sum_{i=1}^n x_i \gtreqless \frac{1}{2}$ is true if and only if $\frac{1}{n}\sum_{i=1}^n (1-x_i) \lesseqgtr \frac{1}{2}$. Hence

$$f(x_1,\ldots,x_n) = \begin{cases} 1 \\ \frac{1}{2} \\ 0 \end{cases} \text{ if and only if } f(1-x_1,\ldots,1-x_n) = \begin{cases} 0 \\ \frac{1}{2} \\ 1, \end{cases}$$

i.e., $f(1-x_1,\ldots,1-x_n) = 1 - f(x_1,\ldots,x_n)$. If moreover $x_i \leq x'_i \leq 1$ for $i \in \{1,\ldots,n\}$, then, by distinguishing the three cases $\frac{1}{n}\sum_{i=1}^n x_i \gtreqless \frac{1}{2}$, it is easy to see that $f(x_1,\ldots,x_n) \leq f(x'_1,\ldots,x'_n)$. Hence, f is self-dual and monotonic, therefore the previously defined rule F_{maj} for KZ-judgment profiles is self-dually systematic and monotonic, and Corollary 1 applies to it. By the conclusion of Corollary 1, F_{maj} yields KZ-consistent Boolean (therefore classically consistent) collective judgments for order compatible KZ-judgment profiles of judgment problems with a closed and internally positive agenda. In particular, this applies to majority aggregation of smooth classical profiles with an odd number of judges.

6 Possibility for logic G

One of the most important t-norm based fuzzy logics is Gödel(-Dummett) logic G (see, e.g., [2]), which features the same semantics for conjunction $\wedge$ and disjunction $\vee$ as Kleene-Zadeh logic KZ. Additionally, we have an implication with the following semantics (valuations v):

$$v(\phi \rightarrow \psi) = \begin{cases} 1 & \text{if } v(\phi) \leq v(\psi) \\ v(\psi) & \text{otherwise.} \end{cases}$$

Negation is defined by $\neg\phi = \phi \rightarrow \bot$, where $v(\bot) = 0$. Moreover, we have $v(\top) = 1$.

To adapt the proof of Theorem 6 to Gödel logic, we have to strengthen order compatibility of profiles and require that the individuals agree for each pair of atoms, whether they should be judged identically or whether one should receive a strictly smaller value than the other.

Definition 20 (Strictly order compatible profiles). *Let $\langle N, \mathcal{A}\rangle$ be a JA problem where $\mathcal{A}$ is closed. The profile $P = \langle J_i \rangle_{i \in N} \in \mathbf{P}$ for $\langle N, \mathcal{A}\rangle$ is strictly order compatible iff there exists an enumeration $\langle p_1, \ldots, p_m \rangle$ of all propositional variables (i.e., non-constant atoms) occurring in the agenda $\mathcal{A}$ such that $\forall i \in N$: $J_i(p_1) \trianglelefteq_1 \cdots \trianglelefteq_{m-1} J_i(p_m)$, where $\trianglelefteq_1, \ldots, \trianglelefteq_{m-1} \in \{=, <\}$.*

Definition 21 (Classically unanimous profiles). *We call a profile $P = \langle J_i \rangle_{i \in N}$ for a judgment aggregation problem $\langle N, \mathcal{A}\rangle$ classically unanimous if the following holds for every $\phi \in \mathcal{A}$: if there is some judge $i \in N$ where $J_i(\phi) = a$ for $a \in \{0, 1\}$, then $J_k(\phi) = a$ for all $k \in N$.*

Lemma 3. *Let $P = \langle J_i \rangle_{i \in N}$ be a strict order compatible and classically unanimous $\mathcal{G}$-judgment profile for the JA problem $\langle N, \mathcal{A}\rangle$, where $\mathcal{A}$ is closed. Then there exists an enumeration $\langle \phi_1, \phi_2, \ldots, \phi_m \rangle$ of all formulas occurring in $\mathcal{A}$, such that*

$$J_i(\phi_1) \trianglelefteq_1 \cdots \trianglelefteq_{m-1} J_i(\phi_m), \text{ where } \trianglelefteq_1, \ldots, \trianglelefteq_{m-1} \in \{=, <\}.$$

for every $i \in N$.[5]

Proof. We proceed by induction on the logical complexity of $\mathcal{A}$, i.e., the total number k of occurrences of logical connectives in the formulas of $\mathcal{A}$. Let $\mathcal{A}_0$ be the set of atoms occurring in $\mathcal{A}$.

If $k = 0$, i.e. if all formulas in $\mathcal{A}$ are atomic, then the claim holds by definition of strict order compatibility and classical unanimity.

In particular, note that if $\bot$ occurs in $\mathcal{A}$ and hence in $\mathcal{A}_0$, then for all $i \in N$ $J_i(\bot) = 0$, since J_i is $\mathcal{G}$-consistent. Hence, due to classical unanimity, we can set $\phi_1 := \bot$, i.e., $\bot$ is the first element of the required enumeration. Similarly, if $\top \in \mathcal{A}$, we set $\phi_m := \top$, i.e., $\top$ is the last element of the enumeration.

If $k > 0$ we have to consider the cases of formulas $\xi = \phi \circ \psi$ with $\circ \in \{\wedge, \vee, \rightarrow\}$. By the induction hypothesis and transitivity either $J_i(\phi) \trianglelefteq J_i(\psi)$ for each $i \in N$ or $J_i(\psi) \trianglelefteq J_i(\phi)$ for each $i \in N$, where $\trianglelefteq \in \{=, <\}$. This implies in particular either

(1) $J_i(\phi) \leq J_i(\psi)$ for all $i \in N$, or

(2) $J_i(\psi) < J_i(\phi)$ for all $i \in N$.

---

[5] As pointed out by an anonymous reviewer, one could strengthen Lemma 3 by only requiring that the derived profile $P_0 = \langle J_1|_{\mathcal{A}_0}, \ldots, J_n|_{\mathcal{A}_0}\rangle$ for $\langle N, \mathcal{A}_0\rangle$, where $\mathcal{A}_0$ denotes the set of atoms of $\mathcal{A}$, is classically unanimous with respect to $\langle N, \mathcal{A}_0\rangle$.

In the former case $J_i(\phi\wedge\psi) = J_i(\phi)$, $J_i(\phi\vee\psi) = J_i(\psi)$ and $J_i(\phi \to \psi) = 1 = J_i(\top)$ for every $i \in N$, while in the latter case, $J_i(\phi\vee\psi) = J_i(\phi)$, $J_i(\phi\wedge\psi) = J_i(\psi) = J_i(\phi \to \psi)$ for every $i \in N$. Therefore, classical unanimity is not violated by $\langle J_i\rangle_{i\in N}$ on $\phi \circ \psi$ since otherwise it would already be violated on some formula (ϕ, ψ or $\top$) from $\mathcal{A}_0$.

For strict order compatibility, we have to consider the following cases:

- $\phi \wedge \psi \in \mathcal{A}$: By the induction hypothesis we have either $J_i(\phi) \leq J_i(\psi)$ for every $i \in N$ or $J_i(\psi) < J_i(\phi)$ for every $i \in N$. Since each judgment J_i is a G-valuation, we have $J_i(\phi \wedge \psi) = \min(J_i(\phi), J_i(\psi))$, which, by the above observation, is either $J_i(\phi)$ for every $i \in N$ or $J_i(\psi)$ for every $i \in N$. Hence, the shared order includes also the judgments for $\phi \wedge \psi$.

- $\phi \vee \psi \in \mathcal{A}$: The argument is analogous to that for $\phi \wedge \psi$: since $J_i(\phi \vee \psi) = \max(J_i(\phi), J_i(\psi))$, the induction hypothesis yields that $J_i(\phi \vee \psi) = J_i(\phi)$ for every $i \in N$ or $J_i(\phi \vee \psi) = J_i(\psi)$ for every $i \in N$. Consequently, the shared order includes $\phi \vee \psi$.

- $\phi \to \psi \in \mathcal{A}$: The argument is similar to the above ones: by the induction hypothesis either (1) $J_i(\phi) \leq J_i(\psi)$ for all $i \in N$, or (2) $J_i(\psi) < J_i(\phi)$ for all $i \in N$. In case (1) $J_i(\phi \to \psi) = 1 = J_i(\top)$ for every $i \in N$ and in case (2) $J_i(\phi \to \psi) = J_i(\psi)$ for every $i \in N$. Consequently, the shared order includes $\phi \to \psi$. □

Theorem 7. *Let $\langle N, \mathcal{A}\rangle$ be a JA problem with a closed agenda $\mathcal{A}$. Then every systematic, strongly monotonic and strongly unanimous aggregation function F yields a G-consistent collective judgment $F(P)$ for every strictly order compatible and classically unanimous G-judgment profile P for $\langle N, \mathcal{A}\rangle$.*

Proof. Let $\langle N, \mathcal{A}\rangle$ be a JA problem with $n := |N|$ and a closed agenda $\mathcal{A}$ and let $P = \langle J_i, \rangle_{i \in N}$ be a strictly order compatible and classically unanimous G-judgment profile for $\langle N, \mathcal{A}\rangle$. Moreover, let F be a systematic, strongly monotonic, and strongly unanimous aggregation function.

We show that $F(P)$ is G-consistent. For this we define $\mathcal{A}_k$ for $k \in \mathbb{N}$ as follows:

- Let $\mathcal{A}_0$ be the set of atoms appearing in $\mathcal{A}$.

- For $k \in \mathbb{N}$, $\mathcal{A}_k$ is defined inductively as

$$\mathcal{A}_{k+1} := \mathcal{A}_k \cup \Big(\mathcal{A} \cap (\{\phi \circ \psi \mid \phi, \psi \in \mathcal{A}_k,\ \circ \in \{\wedge, \vee, \to\}\})\Big).$$

By construction, for $k \in \mathbb{N}$ we have that $\mathcal{A}_k \subseteq \mathcal{A}_{k+1}$, $\mathcal{A}_k \subseteq \mathcal{A}$, and that $\mathcal{A}_k$ is closed. Moreover, for each $k \subset \mathbb{N}$, let $P_k := \langle J_1|_{\mathcal{A}_k}, \ldots, J_n|_{\mathcal{A}_k}\rangle$ be the subprofile of

P restricted to the formulas contained in $\mathcal{A}_k$. I.e., for each agent $i \in N$, P_k contains exactly the judgments of P that judge over the formulas of the restricted agenda $\mathcal{A}_k$. Since P is strictly order compatible and classically unanimous it follows that P_k is also a strict order compatible and classically unanimous G-judgment profile for the G-judgment subproblem $\langle N, \mathcal{A}_k \rangle$. This follows from the fact that for any $k \in \mathbb{N}$, the atoms $\mathcal{A}_0 \subseteq \mathcal{A}_k$ and the individual judgments for each atom in P_k stay the same as in P.

By systematicity, we have $F(P)(\phi) = f(J_1(\phi), \ldots, J_n(\phi))$ for all $\phi \in \mathcal{A}$ and some $f : [0,1]^n \to [0,1]$, which is strongly monotonic by Lemma 2 and closedness of $\mathcal{A}$. This allows us to define a judgment aggregation function F_k on profiles $P' = \langle J'_1, \ldots, J'_n \rangle$ of $\langle N, \mathcal{A}_k \rangle$ by setting $F_k(P')(\phi) := f(J'_1(\phi), \ldots, J'_n(\phi))$ for each $\phi \in \mathcal{A}_k$. In particular, we have $F_k(P_k)(\phi) = f(J_1|_{\mathcal{A}_k}(\phi), \ldots, J_n|_{\mathcal{A}_k}(\phi)) = f(J_1(\phi), \ldots, J_n(\phi)) = F(P)(\phi)$ for $\phi \in \mathcal{A}_k$, i.e., $F_k(P_k) = F(P)|_{\mathcal{A}_k}$.

By induction over k we show that $F_k(P_k)$ is G-consistent for any $k \in \mathbb{N}$. The base case for $k = 0$ is trivial, as all formulas in $\mathcal{A}_0$ are atomic and any assignment of values in $[0, 1]$ to propositional variables constitutes a valuation.

We continue with the induction step from $\mathcal{A}_k$ to $\mathcal{A}_{k+1}$. Let $k \in \mathbb{N}$ and suppose that $F_k(P_k)$ is G-consistent with respect to $\langle N, \mathcal{A}_k \rangle$. We show that then $F_{k+1}(P_{k+1})$ is G-consistent with respect to $\langle N, \mathcal{A}_{k+1} \rangle$.

By definition P_k and P_{k+1} are the judgment subprofiles of P restricted to the formulas in $\mathcal{A}_k$ and $\mathcal{A}_{k+1}$, respectively, with $\mathcal{A}_k \subseteq \mathcal{A}_{k+1}$. So for every formula $\xi \in \mathcal{A}_k$, we have that $F_{k+1}(P_{k+1})(\xi) = F(P)(\xi) = F_k(P_k)(\xi)$, i.e., $F_{k+1}(P_{k+1})|_{\mathcal{A}_k} = F_k(P_k)$. Therefore no inconsistency can arise from formulas in $\mathcal{A}_k$ in the aggregation of P_{k+1} with respect to $\langle N, \mathcal{A}_{k+1} \rangle$.

Next, we show that the unique G-consistent extension of $F_{k+1}(P_{k+1})|_{\mathcal{A}_k}$ to $\mathcal{A}_{k+1}$ coincides with $F_{k+1}(P_{k+1})$. Let $\xi \in \mathcal{A}_{k+1} \setminus \mathcal{A}_k$ be any formula not already contained in $\mathcal{A}_k$. We proceed with a case distinction.

$\xi = \phi \to \psi$: By construction of $\mathcal{A}_{k+1}$, it follows that $\phi, \psi \in \mathcal{A}_k$. Since P_k is a strict order compatible and classically unanimous G-judgment profile for the restricted subproblem $\langle N, \mathcal{A}_k \rangle$, it follows from Lemma 3 that there is an enumeration $\langle \phi_1, \phi_2, \ldots, \phi_m \rangle$ of all formulas occurring in $\mathcal{A}_k$, such that for every $i \in N$:

$$J_i(\phi_1) \trianglelefteq_1 \cdots \trianglelefteq_{m-1} J_i(\phi_m), \text{ where } \trianglelefteq_1, \ldots, \trianglelefteq_{m-1} \in \{=, <\}.$$

Since $\phi, \psi \in \mathcal{A}_k$, by transitivity, we can have the following three cases:

(1) $J_i|_{\mathcal{A}_k}(\phi) < J_i|_{\mathcal{A}_k}(\psi)$ for all $i \in N$
(2) $J_i|_{\mathcal{A}_k}(\phi) = J_i|_{\mathcal{A}_k}(\psi)$ for all $i \in N$

(3) $J_i|_{\mathcal{A}_k}(\psi) < J_i|_{\mathcal{A}_k}(\phi)$ for all $i \in N$

In case (1) and (2), we get for all $i \in N$, $J_i|_{\mathcal{A}_k}(\phi) \leq J_i|_{\mathcal{A}_k}(\psi)$. Since $\phi, \psi \in \mathcal{A}_k$, it holds that $J_i|_{\mathcal{A}_{k+1}}(\phi) = J_i|_{\mathcal{A}_k}(\phi) \leq J_i|_{\mathcal{A}_k}(\psi) = J_i|_{\mathcal{A}_{k+1}}(\psi)$ for every $i \in N$. Thus $J_i|_{\mathcal{A}_{k+1}}(\phi \to \psi) = 1$ for every $i \in N$. By assumption, F is (strongly) unanimous and thus it follows that $F_{k+1}(P_{k+1})(\phi \to \psi) = F(P)(\phi \to \psi) = 1$.

In case (1), since f is a strongly monotonic function and for all $i \in N$, $J_i|_{\mathcal{A}_{k+1}}(\phi) < J_i|_{\mathcal{A}_{k+1}}(\psi)$, we obtain

$$F_{k+1}(P_{k+1})(\phi) = f(J_1|_{\mathcal{A}_{k+1}}(\phi), \ldots, J_n|_{\mathcal{A}_{k+1}}(\phi))$$
$$< f(J_1|_{\mathcal{A}_{k+1}}(\psi), \ldots, J_n|_{\mathcal{A}_{k+1}}(\psi)) = F_{k+1}(P_{k+1})(\psi).$$

In case (2), due to systematicity and since for all $i \in N$, $J_i|_{\mathcal{A}_{k+1}}(\phi) = J_i|_{\mathcal{A}_{k+1}}(\psi)$, we obtain

$$F_{k+1}(P_{k+1})(\phi) = f(J_1|_{\mathcal{A}_{k+1}}(\phi), \ldots, J_n|_{\mathcal{A}_{k+1}}(\phi))$$
$$= f(J_1|_{\mathcal{A}_{k+1}}(\psi), \ldots, J_n|_{\mathcal{A}_{k+1}}(\psi)) = F_{k+1}(P_{k+1})(\psi).$$

Hence, in both cases (1) and (2) $F_{k+1}(P_{k+1})(\phi \to \psi) = 1$ is G-consistent as desired.

In case (3) for all $i \in N$, $J_i|_{\mathcal{A}_{k+1}}(\psi) < J_i|_{\mathcal{A}_{k+1}}(\phi)$ and thus by G-consistency of the individual judgments, for all $i \in N$, $J_i|_{\mathcal{A}_{k+1}}(\phi \to \psi) = J_i|_{\mathcal{A}_{k+1}}(\psi)$. By strong monotonicity of f, we get $F_{k+1}(P_{k+1})(\psi) < F_{k+1}(P_{k+1})(\phi)$ and by systematicity we have

$$F_{k+1}(P_{k+1})(\phi \to \psi) = f(J_1|_{\mathcal{A}_{k+1}}(\phi \to \psi), \ldots, J_n|_{\mathcal{A}_{k+1}}(\phi \to \psi))$$
$$= f(J_1|_{\mathcal{A}_{k+1}}(\psi), \ldots, J_n|_{\mathcal{A}_{k+1}}(\psi)) = F_{k+1}(P_{k+1})(\psi).$$

Hence, with respect to implication, the induction step results in a G-consistent extension.

$\xi = \phi \wedge \psi$: Analogous to $\phi \to \psi$, as a consequence of Lemma 3, we can have the following three cases:

(1) $J_i|_{\mathcal{A}_k}(\phi) < J_i|_{\mathcal{A}_k}(\psi)$ for all $i \in N$

(2) $J_i|_{\mathcal{A}_k}(\phi) = J_i|_{\mathcal{A}_k}(\psi)$ for all $i \in N$

(3) $J_i|_{\mathcal{A}_k}(\psi) < J_i|_{\mathcal{A}_k}(\phi)$ for all $i \in N$

In case (1) and (2), we get for all $i \in N$, $J_i|_{\mathcal{A}_k}(\phi) \leq J_i|_{\mathcal{A}_k}(\psi)$. Since $\phi, \psi \in \mathcal{A}_k$, it holds that $J_i|_{\mathcal{A}_{k+1}}(\phi) = J_i|_{\mathcal{A}_k}(\phi) \leq J_i|_{\mathcal{A}_k}(\psi) = J_i|_{\mathcal{A}_{k+1}}(\psi)$ for every $i \in N$. Thus $J_i|_{\mathcal{A}_{k+1}}(\phi \wedge \psi) = J_i|_{\mathcal{A}_{k+1}}(\phi)$ for every $i \in N$.

In case (1), since f is a strongly monotonic function and for all $i \in N$, $J_i|_{\mathcal{A}_{k+1}}(\phi) < J_i|_{\mathcal{A}_{k+1}}(\psi)$, we obtain

$$F_{k+1}(P_{k+1})(\phi) = f(J_1|_{\mathcal{A}_{k+1}}(\phi), \ldots, J_n|_{\mathcal{A}_{k+1}}(\phi))$$
$$< f(J_1|_{\mathcal{A}_{k+1}}(\psi), \ldots, J_n|_{\mathcal{A}_{k+1}}(\psi)) = F_{k+1}(P_{k+1})(\psi).$$

In case (2), due to systematicity and since for all $i \in N$, $J_i|_{\mathcal{A}_{k+1}}(\phi) = J_i|_{\mathcal{A}_{k+1}}(\psi)$, we obtain

$$F_{k+1}(P_{k+1})(\phi) = f(J_1|_{\mathcal{A}_{k+1}}(\phi), \ldots, J_n|_{\mathcal{A}_{k+1}}(\phi))$$
$$= f(J_1|_{\mathcal{A}_{k+1}}(\psi), \ldots, J_n|_{\mathcal{A}_{k+1}}(\psi)) = F_{k+1}(P_{k+1})(\psi).$$

Hence, in both cases (1) and (2) it follows that the value for the consistent extension is

$$\min(F_{k+1}(P_{k+1})(\phi), F_{k+1}(P_{k+1})(\psi)) = F_{k+1}(P_{k+1})(\phi).$$

Moreover,

$$F_{k+1}(P_{k+1})(\phi \wedge \psi) = f(\langle J_i|_{\mathcal{A}_{k+1}}(\phi \wedge \psi)\rangle_{i \in N})$$
$$= f(\langle J_i|_{\mathcal{A}_{k+1}}(\phi)\rangle_{i \in N}) = F_{k+1}(P_{k+1})(\phi),$$

which is G-consistent.

In case (3) for all $i \in N$, $J_i|_{\mathcal{A}_{k+1}}(\psi) < J_i|_{\mathcal{A}_{k+1}}(\phi)$ and thus by G-consistency of the individual judgments, for all $i \in N$, $J_i|_{\mathcal{A}_{k+1}}(\phi \wedge \psi) = J_i|_{\mathcal{A}_{k+1}}(\psi)$. By strong monotonicity of f, we get $F_{k+1}(P_{k+1})(\psi) < F_{k+1}(P_{k+1})(\phi)$. So it follows that the value for the consistent extension is

$$\min(F_{k+1}(P_{k+1})(\phi), F_{k+1}(P_{k+1})(\psi)) = F_{k+1}(P_{k+1})(\psi).$$

Moreover, by systematicity we have

$$F_{k+1}(P_{k+1})(\phi \wedge \psi) = f(J_1|_{\mathcal{A}_{k+1}}(\phi \wedge \psi), \ldots, J_n|_{\mathcal{A}_{k+1}}(\phi \wedge \psi))$$
$$= f(J_1|_{\mathcal{A}_{k+1}}(\psi), \ldots, J_n|_{\mathcal{A}_{k+1}}(\psi)) = F_{k+1}(P_{k+1})(\psi).$$

$\xi = \phi \vee \psi$: The argument is analogous to that for $\xi = \phi \wedge \psi$.

$\xi = \neg\phi$: Since $\neg\phi = \phi \to \bot$ this case is covered by the case for implication.

We still need to discuss the constant $\bot$. Since the judges evaluate G-consistently, we have $J_i(\bot) = 0$ for all $i \in N$. Using strong unanimity we infer that $F_{k+1}(P_{k+1})(\bot) = F(P)(\bot) = 0$, which is G-consistent. The constant $\top$ does not have to be discussed separately since it is definable as $\top = \bot \to \bot$. However, the argument is analogous to the argument regarding $\bot$, anyway.

This concludes the inductive part of the proof. We further have to show that there is some $k \in \mathbb{N}$ such that for all $l \geq k$, $\mathcal{A}_l = \mathcal{A}$, i.e., that $P_l = P$ and $F(P)$ is G-consistent with respect to $\langle N, \mathcal{A} \rangle$.

This can be done with the same argument as in the proof of Theorem 4. The only difference is that the connective $\circ$ of the composite formula ξ may now be $\wedge$, $\vee$, or $\to$. □

7 Possibility for arbitrary logics via median aggregation

In Sections 5 and 6 we have established possibility results for Kleene-Zadeh logic KZ and Gödel logic G, respectively. These results are specific to the mentioned logics and do not hold, e.g., for Łukasiewicz logic Ł (see Section 4.3). In this section, we will formulate a profile condition that guarantees the consistency of *(propositionwise) median aggregation* for arbitrary many-valued logics.[6] Our result generalizes a well-known possibility result for classical JA by List that shows that the majority judgment is consistent for unidimensionally aligned profiles [14].

Remark 6. *Throughout this section we assume that there are at least two individuals in every JA problem.*

Definition 22 (Median Judgments). *Let $P = \langle J_i \rangle_{i \in N}$ be a profile for a many-valued logic Λ over $[0, 1]$ for some given judgment aggregation problem $\langle N, \mathcal{A} \rangle$. $J^{\mathrm{md}} : \mathcal{A} \to [0, 1]$ is a median judgment for $\phi \in \mathcal{A}$ in P if there exists a bipartition of N into $N = N_\downarrow \cup N_\uparrow$, $N_\downarrow \cap N_\uparrow = \emptyset$, such that $|N_\downarrow| = |N_\uparrow|$ or $|N_\downarrow| = |N_\uparrow|+1$ or $|N_\downarrow| = |N_\uparrow|-1$ and the following holds: $\forall i \in N_\downarrow : J_i(\phi) \leq J^{\mathrm{md}}(\phi)$ and $\forall i \in N_\uparrow : J_i(\phi) \geq J^{\mathrm{md}}(\phi)$. If $J^{\mathrm{md}}(\phi)$ coincides with J_i for some $i \in N$, we call i a median judge with respect to J^{md} for ϕ in P.*

---

[6]Propositionwise median aggregation, as introduced here, should not be confused with the 'median rule' for classical judgment aggregation investigated by Nehring and Pivato in [18]. The latter rule selects a consistent classical collective judgment that minimizes the average distance to individual (classical) judgments. Even if one restricts attention to classical judgments, our (propositionwise) median aggregation is different from median aggregation in the sense of [18]. However, the potential power and usefulness of our median rule only emerges in the many-valued setting.

J^{md} *is a* (propositionwise) *median judgment for P and $\mathcal{A}$ if J^{md} is a median judgment for every $\phi \in \mathcal{A}$ in P. If J^{md} coincides with J_i for every $\phi \in \mathcal{A}$ then i is a median judge in P for $\mathcal{A}$.*

We emphasize that, for any profile, an appropriate bipartition and hence a median judgment exists for each $\phi \in \mathcal{A}$. But for different agenda items $\phi, \psi \in \mathcal{A}$ the bipartitions do not have to agree. Therefore median judgments for the whole agenda are not consistent in general. However, if, for a profile P, there exists an agent who is a median judge for every agenda item and hence also a median judge in P, then the corresponding median judgment is consistent by definition.

Definition 23 (Median Aggregation Functions). *Let $\langle N, \mathcal{A} \rangle$ be a Λ-JA problem for a many-valued logic Λ over $[0,1]$. Then a function $F^{\mathrm{md}} : \mathbf{P} \to [0,1]^{\mathcal{A}}$ is a* (propositionwise) *median aggregation function if, for every $P \in \mathbf{P}$ and every $\phi \in \mathcal{A}$, $F^{\mathrm{md}}(P)(\phi)$ is a median judgment for ϕ in P.*

As indicated above, median judgments exist for each proposition $\phi \in \mathcal{A}$ in every profile $\langle J_i \rangle_{i \in N}$. Moreover, for each $\phi \in \mathcal{A}$ there is a also a corresponding median judge. However, in general, there is no median judge for the whole agenda $\mathcal{A}$ and hence the median aggregation function may yield an inconsistent output. (Note that this fact is independent from the underlying logic Λ.)

If one orders the agents N in such a way that the corresponding judgment values $J_i(\phi)$ appear in either ascending or descending order, then both $J_m(\phi)$ and $J_{m+1}(\phi)$, are median judgments (and hence m and $m+1$ are median judges) for ϕ if $|N| = 2m$. If $|N| = 2m+1$ one even obtains three median judgments: $J_m(\phi)$, $J_{m+1}(\phi)$ and $J_{m+2}(\phi)$. This implies that a median judgment is only useful if there are more than three individuals.[7] On the other hand, if there are a lot of agents and if the individual values attached to propositions are widespread, then median judgments may well amount to representative aggregations. One may argue that determining the average values, rather than picking a median value, is even more informative. However, note that for every proposition ϕ there exists only a single average value, whereas, in general, there are entire intervals of possible median judgments for ϕ. In the above scenario, where m and $m+1$ and, for the case $|N| = 2m+1$, also $m+2$ are median judges for ϕ, we may set $J^{\mathrm{md}}(\phi) = x$ for any $x \in [J_m(\phi), J_{m+1}(\phi)]$ or $x \in [J_m(\phi), J_{m+2}(\phi)]$, respectively. Thus uncountably many different median judgments may arise. As a consequence, median aggregation is more flexible than average aggregation in general.

---

[7]In particular, for the special case of the discursive dilemma (see Section 3.2), every individual judgment constitutes a median judgment for the given profile. While this hardly amounts to a resolution of the dilemma, it reflects the fact that any of the three individual judgments seems equally representative in this case.

Example 2. *Consider the following profile P for a JA-problem $\langle N, \mathcal{A}\rangle$, where $N = \{1,\ldots,6\}$ and $\mathcal{A} = \{\phi_1, \phi_2, \phi_3, \phi_4\}$.*

	ϕ_1	ϕ_2	ϕ_3	ϕ_4
$J_1:$	0	1	1	0
$J_2:$	0.2	0.7	0.7	0.2
$J_3:$	0.3	0.5	0.5	0.3
$J_4:$	0.4	0.4	0.4	0.4
$J_5:$	0.8	0.3	0.8	0.3
$J_6:$	1	0	1	0

Note that for every $i \in N$ we have $J_i(\phi_3) = \max(J_i(\phi_1), J_i(\phi_2))$ and $J_i(\phi_4) = \min(J_i(\phi_1), J_i(\phi_2))$. Hence, we have Λ-consistent individual judgments if $\phi_3 = \phi_1 \vee \phi_2$ and $\phi_4 = \phi_1 \wedge \phi_2$ if Λ is a logic where disjunction is modeled by maximum and conjunction by minimum, like, for example, in Kleene-Zadeh logic KZ or Gödel logic G.

Any median judgment J^{md} for P and $\mathcal{A}$ has to satisfy the following constraints: $J^{\mathrm{md}}(\phi_1) \in [0.3, 0.4]$, $J^{\mathrm{md}}(\phi_2) \in [0.4, 0.5]$, $J^{\mathrm{md}}(\phi_3) \in [0.7, 0.8]$, and $J^{\mathrm{md}}(\phi_4) \in [0.2, 0.3]$. If we assume that $\phi_3 = \phi_1 \vee \phi_2$ and $\phi_4 = \phi_1 \wedge \phi_2$, then there is no KZ- or G-consistent median judgment for P and $\mathcal{A}$, since the possible median values for $\phi_3 = \phi_1 \vee \phi_2$ are strictly higher than those for ϕ_1 and ϕ_2. However, if we delete ϕ_3 from the agenda, we can satisfy the constraints. Indeed, individual 3 is a median judge for $\mathcal{A} - \{\phi_3\}$.

If we change $J_2(\phi_2)$ and $J_2(\phi_3)$ to 0.5 then we obtain the constraint $J^{\mathrm{md}}(\phi_3) \in [0.5, 0.8]$ for $\phi_3 = \phi_1 \vee \phi_2$. In that case, individual 3 is a median judge for the whole agenda $\mathcal{A}$.

Median judgments are related to majority judgments. For classical JA, where $J_i(\phi) = 0$ or $J_i(\phi) = 1$ for every $i \in N$ and every $\phi \in \mathcal{A}$, median judgments are non-discriminating, i.e. $J^{\mathrm{md}}(\phi)$ can be 0 as well as 1, for a proposition ϕ, if and only if (almost) as many judges evaluate ϕ to 1 and to 0, respectively. More precisely, let $n^+(\phi) = |\{i \in N \mid J_i(\phi) = 1\}|$ and $n^-(\phi) = |\{i \in N \mid J_i(\phi) = 0\}|$. Then $J^{\mathrm{md}}(\phi)$ can take both values if $n^+(\phi) = n^-(\phi)$ (for even $|N|$) or $|n^+(\phi) - n^-(\phi)| = 1$ (for odd $|N|$). In all other cases, the median judgment coincides with the majority judgment.

Also in the many-valued scenario, median aggregation has attractive features. In general, median aggregation functions are non-dictatorial. By this we mean the following: If there are more than 3 agents and if the agenda contains at least one formula that can take any value in $[0, 1]$, then, for any specific agent i, there exists a profile for which i is not a median judge. Again, this observation does not depend on the underlying many-valued logic.

Informally, one may say that median judgments keep a balance between sets of individuals that value a given proposition at least as high or at least as low, respectively, as some intermediate value. Median aggregation returns such an intermediate *median value* as the collective value for each proposition of the agenda. In contrast to the *average* of the individuals' values, median values are not unique, as we have seen above. The possibility to choose between different median values implies that one can, in principle, define median aggregation functions that are neither anonymous nor systematic. However, if one keeps the choice of the median value independent from the identity of individuals and uniform for all propositions, then one obtains an anonymous and systematic median aggregation function. For example, we may stipulate that the aggregation function always assigns the lowest (or highest) possible median judgment value to a given proposition.

As we have seen in Example 2, median aggregation, just like average aggregation, does not yield consistent collective judgments in general. However, the following restriction on profiles, which generalizes a notion defined in [14] for classical JA, provides a sufficient condition for the existence of a median judge for the whole agenda.

Definition 24 (Unidimensional Alignment). *Let $P = \langle J_i \rangle_{i \in N} \in \mathbf{P}$ be a profile for an Λ-JA problem $\langle N, \mathcal{A} \rangle$. P is* unidimensionally aligned *with respect to a (strict) linear ordering $\prec$ of the individuals N if, for every $\phi \in \mathcal{A}$ one of the following holds.*

(1) $\forall i, j \in N \ i \prec j \Rightarrow J_i(\phi) \leq J_j(\phi)$, or

(2) $\forall i, j \in N \ i \prec j \Rightarrow J_i(\phi) \geq J_j(\phi)$.

In case (1) ϕ is upwards aligned *in P with respect to $\prec$. In case (2) ϕ is* downwards aligned *in P with respect to $\prec$.*

A profile P is unidimensionally alignable *if there exists a linear ordering $\prec$ of N such that P is unidimensionally aligned with respect to $\prec$.*

Note that in contrast to (strictly) order compatible profiles (see Sections 5 and 6), unidimensionally aligned profiles are ordered along individuals, not along propositions.

Example 3. *We revisit the profile P of Example 2. Proposition ϕ_1 is upwards aligned and proposition ϕ_2 is downwards aligned in P with respect to the natural order relation ($<$) on $N = \{1, \ldots, 6\}$. But ϕ_3 and ϕ_4 are not unidimensionally aligned with respect to $<$. We can of course, downwards or upwards align those propositions individually with respect to some other ordering of N. However, there is no ordering $\prec$ of N such that all four propositions in $\mathcal{A}$ are unidimensionally* aligned with respect to $\prec$. Hence P is not unidimensionally alignable. In fact, even if we delete either ϕ_3 or ϕ_4 from $\mathcal{A}$ and correspondingly from P, the two resulting profiles are not unidimensionally alignable. On the other hand, the profile restricted to the agenda $\mathcal{A}' = \{\phi_1, \phi_2\}$ is already unidimensionally aligned.

Theorem 8. *Given a JA problem $\langle N, \mathcal{A} \rangle$ with respect to any many-valued logic Λ, there exists a median judge in every unidimensionally alignable profile for $\mathcal{A}$.*

Proof. Let $P = \langle J_i \rangle_{i \in N} \in \mathbf{P}$ be a unidimensionally alignable profile for the JA problem $\langle N, \mathcal{A} \rangle$. By Definition 24, there is a linear ordering $\prec$ of N such that P is unidimensionally aligned with respect to $\prec$. In particular, every $\phi \in \mathcal{A}$ is either upwards aligned or downwards aligned in P with respect to $\prec$.

We distinguish two cases, according to whether $|N|$ is odd or even.

$|N|$ *is odd*: Since N is linearly ordered by $\prec$, there is a midpoint $m \in N$ with respect to $\prec$. This means that we can partition N into $N_1 = \{i \mid i \prec m\}$ and $N_2 = \{i \mid m \prec i\} \cup \{m\}$ such that $|N_1| = |N_2| - 1$. Note that, since $\prec$ is a strict linear ordering we have $N = N_1 \cup N_2$ and $N_1 \cap N_2 = \emptyset$. If $\phi \in \mathcal{A}$ is upwards aligned in P with respect to $\prec$ then $\forall i \in N_1\ J_i(\phi) \leq J_m(\phi)$ and $\forall i \in N_2\ J_i(\phi) \geq J_m(\phi)$. Hence J_m is a median judgment for ϕ according to Definition 22, where N_1 instantiates $N_\downarrow$ and N_2 instantiates $N_\uparrow$. If, on the other hand, ϕ is downwards aligned in P with respect to $\prec$ then $\forall i \in N_1\ J_i(\phi) \geq J_m(\phi)$ and $\forall i \in N_2\ J_i(\phi) \leq J_m(\phi)$. Again J_m is a median judgment for ϕ, where now N_2 instantiates $N_\downarrow$ and N_1 instantiates $N_\uparrow$ in the defining condition. To sum up, J_m is a median judgment for every $\phi \in \mathcal{A}$, and hence m is a median judge in the profile P.

$|N|$ *is even*: The argument is analogous to the previous case. We now obtain two adjacent midpoints m and m' with respect to $\prec$. We partition N into $N_1 = \{i \mid i \prec m\} \cup \{m\}$ and $N_2 = \{i \mid m' \prec i\} \cup \{m'\}$ and obtain that both, m and m' are median judges for P. $\square$

Corollary 2. *For every JA problem $\langle N, \mathcal{A} \rangle$ there is a median aggregation function F that yields a Λ-consistent collective judgment for every unidimensionally alignable profile for $\langle N, \mathcal{A} \rangle$.*

Proof. If $P = \langle J_i \rangle_{i \in N}$ is a unidimensionally alignable profile for $\langle N, \mathcal{A} \rangle$, then Theorem 8 entails that there is a median judge $m \in N$ in P for $\mathcal{A}$. But, by definition, the judgment of every agent in N is Λ-consistent. Hence we obtain a median aggregation function F with the required property if we set $F(P) = J_m$ whenever P is a unidimensionally alignable profile, and let $F(P)$ be any median judgment for P and $\mathcal{A}$, otherwise. $\square$

To summarize, if the agenda is unidimensionally aligned, then a median judgment coincides with the judgment of one of the individuals (the median judge) for the entire agenda and is therefore Λ-consistent by definition.

We have already indicated above how average aggregation compares with median aggregation, introduced here. We emphasize that median aggregation leaves more room for consistent collective judgments than average aggregation. For example, for any classical (0/1-valued) profile where $|N| = 3$, each individual is a median judge and thus provides a consistent collective median judgment. In contrast, the doctrinal paradox (see Section 2.1) exhibits a classical profile that cannot be consistently aggregated with the average rule, even if we allow intermediate values in the collective judgment. The greater 'liberality' of median aggregation is related to the fact that it is based only on comparisons of individual judgments, whereas calculating the average requires arithmetic operations to be applied to individual judgments.

8 Conclusion

Motivated by the fact that judgment aggregation for many-valued logics has so far focused on impossibility theorems [7, 11], we have provided some positive (possibility) results. We have identified conditions on judgment profiles that guarantee consistent collective judgments by averaging for certain types of agendas using Kleene-Zadeh logic KZ and Łukasiewicz logic Ł, respectively. The result for KZ has then been generalized to a wider family of aggregation functions. Additionally, we showed a similar result for Gödel-Dummett logic G, which also includes implication, a connective missing in KZ. Finally, we introduced median aggregation and presented a possibility result that applies to arbitrary many-valued logics. The corresponding profile condition generalizes List's concept of unidimensionally alignable profiles [14] from classical to many-valued judgments.

The mentioned results should be considered as first steps within a wider research program that systematically surveys the frontier region separating possibilities and impossibilities for consistent judgment aggregation in a many-valued setting.

Acknowledgment

We are grateful to an anonymous reviewer for an exceptionally detailed and thoughtful report on an earlier version of the paper. The report led to corrections of many faulty and ambiguous statements that have crept into the conference papers [8] and [21] on which the present paper is based.

References

[1] Stefano Aguzzoli, Brunella Gerla, and Vincenzo Marra. Algebras of fuzzy sets in logics based on continuous triangular norms. In *Proceedings of the 10th European Conference on Symbolic and Quantitative Approaches to Reasoning with Uncertainty (ECSQARU 2009)*, pages 875–886. Springer, 2009.

[2] Matthias Baaz and Norbert Preining. Gödel-Dummett logics. In Petr Cintula, Petr Hájek, and Carles Noguera, editors, *Handbook of Mathematical Fuzzy Logic - Volume 2*, pages 585–625. College Publications, 2015.

[3] Franz Dietrich. A generalised model of judgment aggregation. *Social Choice and Welfare*, 28:529–565, 2007.

[4] Franz Dietrich and Christian List. Arrow's theorem in judgment aggregation. *Social Choice and Welfare*, 29(1):19–33, 2007.

[5] Franz Dietrich and Christian List. From degrees of belief to binary beliefs: Lessons from judgment-aggregation theory. *The Journal of Philosophy*, 115(5):25–270, 2018.

[6] Elad Dokow and Ron Holzman. Aggregation of binary evaluations with abstentions. *Journal of Economic Theory*, 145(2):544–561, 2010.

[7] María Esteban, Alessandra Palmigiano, and Zhiguang Zhao. An abstract algebraic logic view on judgment aggregation. In *Proceedings of the International Workshop on Logic, Rationality and Interaction*, pages 77–89. Springer, 2015.

[8] Christian G. Fermüller. Some consistency criteria for many-valued judgment aggregation. In *53rd IEEE International Symposium on Multiple-Valued Logic, ISMVL 2023, Matsue, Japan, May 22-24, 2023*, pages 215–220. IEEE, 2023.

[9] Davide Grossi and Gabriella Pigozzi. *Judgment Aggregation: A Primer*. Morgan & Claypool Publishers, 2014.

[10] Petr Hájek. *Metamathematics of Fuzzy Logic*, volume 4 of *Trends in Logic*. Kluwer, 1998.

[11] Frederik S. Herzberg. Universal algebra for general aggregation theory: Many-valued propositional-attitude aggregators as MV-homomorphisms. *Journal of Logic and Computation*, 25(3):965–977, 2013.

[12] Lewis A. Kornhauser and Lawrence G. Sager. The one and the many: Adjudication in collegial courts. *California Law Review*, 81(1):1–59, 1993.

[13] Jérôme Lang and Marija Slavkovik. Judgment aggregation rules and voting rules. In *Proceedings of the International Conference on Algorithmic Decision Theory*, pages 230–243. Springer, 2013.

[14] Christian List. A possibility theorem on aggregation over multiple interconnected propositions. *Mathematical Social Sciences*, 45(1):1–13, 2003.

[15] Christian List. Social Choice Theory. In Edward N. Zalta, editor, *The Stanford Encyclopedia of Philosophy*. Metaphysics Research Lab, Stanford University, Spring 2022 edition, 2022.

[16] Christian List and Philip Pettit. Aggregating sets of judgments: An impossibility result.

Economics and Philosophy, 18:89 – 110, 2002.

[17] Philippe Mongin. Judgment aggregation. In Sven Ove Hansson, Vincent F. Hendricks, and Esther Michelsen Kjeldahl, editors, *Introduction to Formal Philosophy*, chapter 38, pages 705–720. Springer, 2018.

[18] Klaus Nehring and Marcus Pivato. The median rule in judgement aggregation. *Economic Theory*, 73(4):1051–1100, 2022.

[19] Hung T. Nguyen, Carol Walker, and Elbert A. Walker. *A First Course in Fuzzy Logic*. CRC Press, fourth edition, 2018.

[20] Philip Pettit. Deliberative democracy and the discursive dilemma. *Philosophical Issues*, 11(1):268–299, 2001.

[21] Sebastian Uhl and Christian G. Fermüller. Many-valued judgment aggregation – some new possibility results. In Van-Nam Huynh, Bac Le, Katsuhiro Honda, Masahiro Inuiguchi, and Youji Kohda, editors, *Integrated Uncertainty in Knowledge Modelling and Decision Making*, pages 3–14, Cham, 2023. Springer Nature Switzerland.

www.ingramcontent.com/pod-product-compliance
Lightning Source LLC
Chambersburg PA
CBHW080453170426
43196CB00016B/2787